ELECTRONS IN METALS

A Short Guide to the Fermi Surface

By

J. M. ZIMAN, F.R.S.

University Lecturer in Physics, and Fellow of King's College, Cambridge
(Now at H. H. Wills Physics Laboratory, Royal Fort, Bristol, 8.)

TAYLOR & FRANCIS LTD
10-14 MACKLIN STREET, LONDON, WC2B 5NF

1970

First published 1963 by Taylor & Francis Ltd., 10/14 Macklin Street, London, WC2B 5NF
(reprinted from a series of articles which appeared during 1962 in *Contemporary Physics*)

Reprinted 1964
Reprinted 1966
Reprinted 1970

© 1970 Taylor & Francis Ltd

All rights reserved. No part of this publication may be reproduced, stored in a retrieval system or transmitted, in any form or by any means, electronic, mechanical, photocopying, recording or otherwise, without the prior permission of the Copyright owner.

Printed photo offset by Warren & Son Ltd (a member of the Taylor & Francis Group)
The Wykeham Press, Winchester, Hampshire.

SBN 85066 004 1

Distributed in the United States of America and its territories by Barnes & Noble Inc.,
105 Fifth Avenue, New York, New York 10003.

CONTENTS

		Page
Part I.	The Electron Gas	1
Part II.	Bands and Zones	16
Part III.	Dynamics of Bloch Electrons and the Calculation of Band Structure	29
Part IV.	The Properties of Real Metals	43
Part V.	Gauging the Fermi Surface	57
	Acknowledgements	75
	References	75

ELECTRONS IN METALS

A Short Guide to the Fermi Surface

by J. M. ZIMAN

University Lecturer in Physics, and Fellow of King's College, Cambridge

Part I. The Electron Gas

'*Either be wholly slaves or wholly free.*'—Dryden

1. INTRODUCTION

Browsing in books, review articles and conference reports, one sometimes nowadays comes across pictures like fig. 1. This peculiar object, which might be mistaken for a piece of modern sculpture, is labelled 'The Fermi surface of magnesium'. What does it mean, to say that magnesium has a Fermi surface,

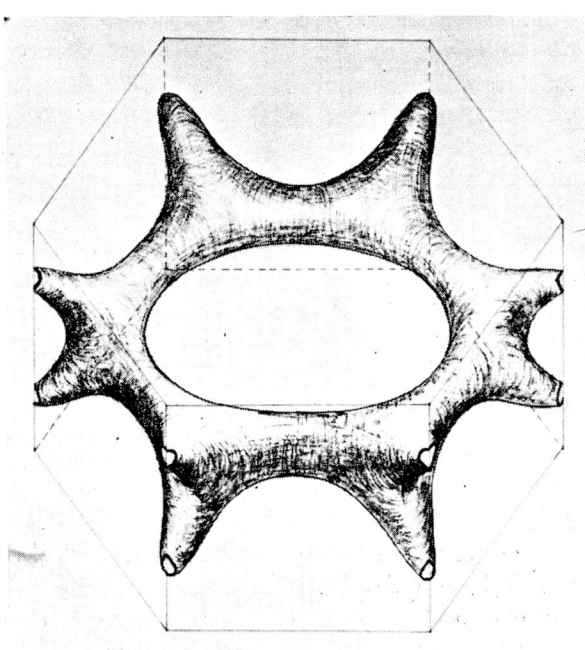

Fig. 1. The Fermi surface of magnesium, or Leo Falicov's Monster.

and how is it determined? This is the first of a series of articles in which an attempt is made to answer these questions.

The Fermi surface is a mathematical construction related to the dynamical properties of the conduction electrons in a metal. To understand its significance we must first accept the idea that these electrons form a gas obeying Fermi-Dirac statistics. This is dealt with in the present article. In the second article we consider the effect on this gas of the regular crystalline lattice of the ions of the metal amongst which the electrons move. We show that this creates gaps in the range of energies allowed to the electrons, and we learn how to construct the Brillouin zone—a polyhedron in momentum space whose boundaries define the positions of the energy gaps.

The third article is devoted to a study of the dynamical properties of the electrons in the allowed energy bands, under the influence of electric and magnetic fields. We learn here how these properties can be derived from the shape of the Fermi surface, and we study the problem of calculating the shape of the Fermi surface from first principles. We shall find that the idea that the electrons are nearly free can be justified mathematically, in spite of the very strong interactions between electrons and ions, and between the conduction electrons themselves. In the fourth article we look at real metals, especially the alkali metals, and show how such properties as electrical conductivity and thermoelectric power are determined by the shape of the Fermi surface; this article also gives an account of the anomalous skin effect, which provided the first detailed picture of the complicated shape of the Fermi surface in copper.

In the final article, we look at other methods for the experimental study of the Fermi surface. These methods all depend on the use of strong magnetic field which cause the electrons to move in helical paths in the metal. These paths have their representation in momentum space as orbits on the Fermi surface, and give rise to characteristic phenomena such as cyclotron resonance, magneto-resistance, the de Haas–van Alphen effect, the magneto-acoustic effect, etc. By studying the variation of these effects as one varies the direction of the magnetic field, one can map out the Fermi surface, and make a model of it for each metal.

These articles are only travellers' tales, not a Baedeker. I have only tried to make the general theory plausible; there are many admirable books which deal with the subject in all degrees of rigour and which are listed at the end of this chapter. For this reason, no specific references to the original literature will be given in the main text; it may be assumed that these can all be discovered in the books which I have listed.

2. Mobile electrons

Consider sodium. We all know that it is a monovalent element because it has just one electron outside a closed shell. This electron is very loosely bound, and may easily be lost, leaving a sodium ion. Almost all the ordinary chemistry of sodium refers to this ion, in water, in crystals such as NaCl, etc.

Now put two neutral sodium atoms together. They will begin to interact with one another. Atom A may attract the valence electron from B and ionize it. Or, both valence electrons may prefer to go to B, leaving A temporarily bare. If the two nuclei (bearing their closed shells) come near enough together, it becomes possible for both the electrons to circle about both ions together, just as if we had a doubly-charged nucleus (e.g. He) with two electrons in orbit around it. In fact this is a rather favourable arrangement, because the two electrons form almost a closed atomic sub-shell; the Na_2 molecule is normal in the vapour.

Now let us pack together more and more sodium atoms, as in a solid or a liquid. Each neighbouring pair of electrons would like to form a molecule and share the electrons. But each atom has eight neighbours and has only one electron to be shared with all of them. So this electron is sent on a very complex course, visiting each neighbour in turn. The other eight electrons from these neighbours also travel round the complex, each of them spending about one eighth of its time with the central atom of our little group. Then, since this is only one cell of a complete crystal lattice, other electrons will come in from further cells—and so on. In other words, instead of having a cosy arrangement in which each ion has its ' own ' electron bound tightly to it, we have a sort of communist society in which all the ions possess all the electrons in common, and the electrons can move freely from one ion to another.

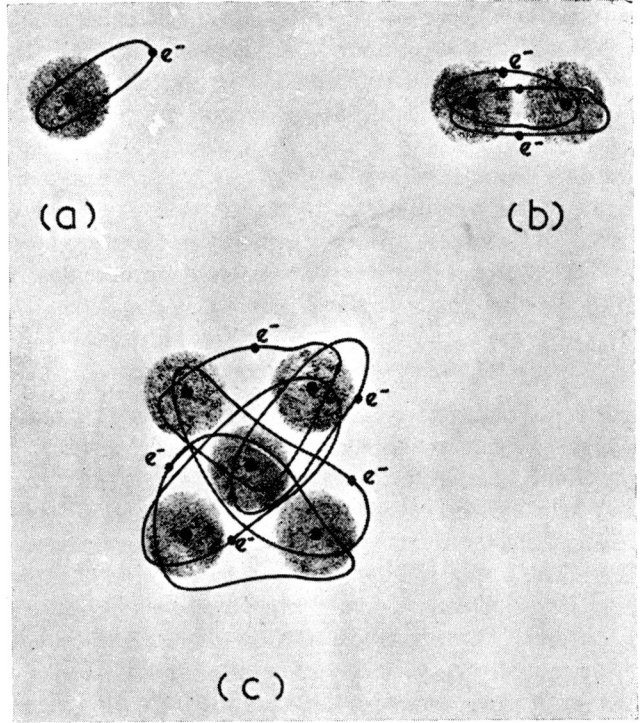

Fig. 2. (a) Sodium atom; (b) Sodium molecule; (c) Sodium metal.

This idea is quite enough to explain, in general terms, many characteristic properties of metals, most of which stem from the fact that metals are good conductors of electricity. Obviously, if the electrons are very mobile, we may apply an electric field to draw them out of one side of the crystal—and go on drawing them out so long as we feed them back again on the other side to keep the whole crystal electrically neutral. That is, we can easily pass an electric current through the solid.

Contrast this behaviour with what happens in an insulator, such as NaCl. Here the valence electron of the sodium has been swallowed by the chlorine atom, giving an Na^+ ion and a Cl^- ion, both with complete closed shells. To make a current flow, we must either make the chlorine ion regurgitate the electron—a process costing many electron volts in energy—or we must force a whole ion to move through the lattice—a very slow process, only possible in practice if the crystal structure is imperfect.

3. Jellium

It is typical of modern physicists that they will erect skyscrapers of theory upon the slender foundations of outrageously simplified models. We have an array of ions, each with charge enough to bind an electron if left to itself, and an equal number of electrons which seem to be able to move around a bit amongst the ions. So we take the bit between our teeth, and assume that the ions are not there at all! Or, rather we smear them out into a fixed uniform background of positive charge—a sort of jelly—in which the electrons move quite freely. We need this jelly to make our system electrically neutral, so that the electrons will not be driven explosively apart by their coulomb repulsion. It is sometimes instructive to think about the effects of elastic waves in this medium, but generally speaking it can be ignored. Thus the properties of *jellium* are those of a *free electron gas*.

On the face of it this model seems wildly unrealistic. How can we simply ignore those strong local forces, varying so rapidly in the neighbourhood of each ion? In Part III we shall see that it is a much better representation of the metal than one would have thought at first.

4. The Wiedemann–Franz law

We saw that high electrical conductivity is the basic 'metallic' property. Naturally we wish to calculate this for our free electron gas. That is easy; the conductivity would be infinite. But then if we think of jellium, we remember that the positive jelly will be subject to fluctuations of density due to thermal vibrations. A local condensation of the jelly (i.e. a local increase of average packing density of the ions) will look like a local positive charge, and this will scatter the conduction electrons. Again, there may be an impurity in the metal, an element of different valency whose ion cannot be smeared into the background. Or, again, there may be a vacant site, or an interstitial atom, or a dislocation line, or a grain boundary, where we can no longer suppose the medium to be homogeneous and smooth. All such objects will scatter the electrons in the gas[†].

To allow for such effects let us introduce a *relaxation time*, τ, such that an electron is only free for τ seconds (on the average), before it is scattered. Now apply an electrical field, **E**. The force on the electron will be $e\mathbf{E}$, and it will be accelerated. In τ seconds it will acquire a drift velocity

$$\mathbf{u} \sim e\mathbf{E}\tau/m$$

in the direction of the field. After each collision with an impurity, etc., this velocity will be lost, or so changed in direction as to be effectively zero. If there are n electrons per unit volume, each carrying the charge e, this drift motion will be equivalent to an electric current density

$$\mathbf{J} \sim ne\mathbf{u}. \tag{1}$$

[†] But the actual ions of the lattice do not scatter the electrons; they alter their dynamical properties. This will appear in Parts II and III.

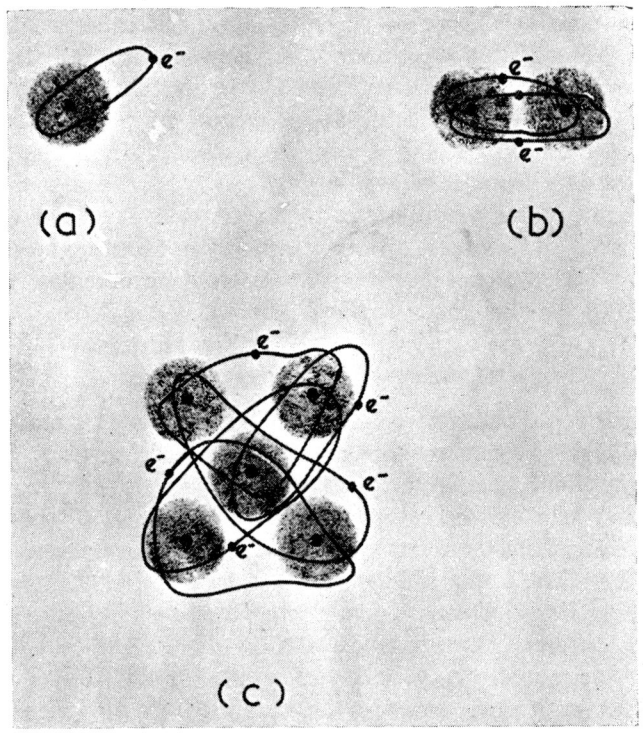

In other words the electrical conductivity of our electron gas can be written

$$\sigma = J/E \sim ne^2\tau/m. \tag{2}$$

This very simple and important formula tells us what we might have guessed, that the conductivity of a metal is proportional to the number of free electrons per unit volume. But to compare this with experiment we need a theory for the relaxation time τ, and this is far above our present level of argument. Nevertheless there is one very instructive comparison that can easily be made.

Think what happens if, instead of an electric field, we establish a temperature gradient, grad T, along our specimen. At temperature T, an electron has energy $\frac{3}{2}kT$, where k is the Boltzmann constant. An electron in a 'hot' region—temperature $T + \delta T$ say—will have excess energy $\frac{3}{2}k\delta T$, and will tend to diffuse into a 'colder' region. This excess energy changes as we go along the wire, just as if there were a field of potential acting along it. It is as if there were a force of magnitude $\frac{3}{2}k$ (grad T) acting on each electron. This thermodynamic force gives rise to a drift of electrons down the wire, controlled, as in the electrical case, by collisions with imperfections, etc. The average drift velocity comes out by the same argument:

$$\mathbf{u} \sim \tfrac{3}{2}k(\text{grad } T)\,\tau/m.$$

Each electron carries heat $\frac{3}{2}kT$ with it. The heat current will thus be

$$\mathbf{U} \sim (\tfrac{3}{2})^2 nk^2 T\,\mathbf{u} \tag{3}$$

as if there were a *thermal conductivity* of the electron gas:

$$\kappa = U/(\text{grad } T) \sim (\tfrac{3}{2})^2 nk^2 T\tau/m. \tag{4}$$

The beauty of these formulae is that we may eliminate the unknown relaxation time, and discover the following general relation:

$$\kappa/\sigma T \sim (\tfrac{3}{2})^2 k^2/e^2. \tag{5}$$

This tells us that the electrical and thermal conductivities of a free electron gas are not independent of one another: it has been known as an empirical law since it was discovered by Wiedemann and Franz in 1853. Our derivation is rather informal (as indicated by the use of twiddles instead of equality signs) but it turns out to be quite a sound result. The strict derivation, in the full panoply of statistical quantum mechanics, only introduces a numerical factor $\pi^2/3$ instead of $(3/2)^2$ but tells us that the law is not true if the electron is scattered with change of energy.

We now see that high thermal conductivity is also characteristic of metals, being associated with the high electrical conductivity of the electron gas. The experimental confirmation of this relation tells us that the same carriers are responsible for both phenomena.

5. The Hall effect

There is another transport effect which can be discussed in terms of this simple classical model. Suppose that we have a strip of metal with an electric current flowing uniformly along it. We apply a magnetic field at right angles to the strip. The electrons, being charges moving in a magnetic field, will be deflected to one side. We might think that this would impede the flow of current. But a charge will then build up along the edge of the strip. Further electrons approaching the edge will be deflected back. We soon achieve a stationary state, in which the deflection caused by the magnetic field is balanced by the electric field of the system of charges along the edges of the strip, and current flows as before. This electric field can be detected as a potential difference at right angles to the current direction and to the magnetic field, and is called the *Hall field*.

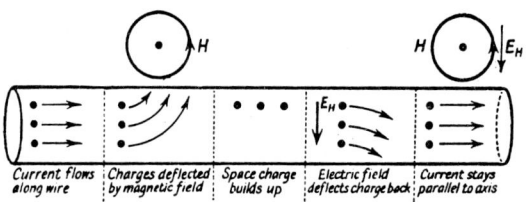

Fig. 3. The Hall effect.

To calculate this effect is easy. Each electron moves with drift velocity **u**, so that, by the standard Lorentz formula of electrodynamics, the magnetic field exerts a force

$$\mathbf{F} = \frac{e}{c}\mathbf{u} \wedge \mathbf{H}. \tag{6}$$

This will be exactly balanced if the Hall field is of strength

$$e\mathbf{E}_H = -\mathbf{F} = -\frac{e}{c}\mathbf{u} \wedge \mathbf{H}. \tag{7}$$

We can express this in terms of the electric current density, $\mathbf{J} = ne\mathbf{u}$, by multiplying through by the density of electrons in the specimen. Thus

$$\mathbf{E}_H = +\frac{1}{nec}\mathbf{H} \wedge \mathbf{J}. \tag{8}$$

In the standard configuration where **J** and **H** are deliberately constrained to be at right angles, we can say that the Hall voltage is proportional to the strength of the magnetic field, and to the magnitude of the current, with the *Hall coefficient*

$$R = \frac{1}{nec}. \tag{9}$$

This derivation of the formula is again rather crude, but the result is correct for a free-electron gas, classical or quantal. It is interesting because we can use it to discover the sign of the carriers, and their density. The Hall effect is linear in e, and should therefore be negative for electrons (it is a good exercise in mental ping-pong to visualize fig. 3 with the charges reversed in sign!). Most metals do have 'normal' Hall coefficients, but there are some quite ordinary metals (Pb, Zn) which behave as if the carriers were positively charged. Even in the normal case, we do not always find that the density of electrons is the same as the total number of valence electrons that can be put into the pool.

These anomalies cannot be explained within the free-electron model; we shall see later how they arise through the interaction of the electrons with the ionic lattice.

Another anomaly may be noted. The Hall field exactly balances the magnetic field, so that the electrons are, in the end, subject to no transverse forces. Thus the conductivity of the metal *measured along the current direction* does not depend on the magnetic field. Experimentally, however, all metals show a *transverse magnetoresistance* (increase of electrical resistance of a wire placed transverse to a magnetic field) which would not occur in a free electron gas.

6. Specific heat and thermoelectric effect

The ionic lattice, with N ions per unit volume, will have a specific heat, at high temperatures, $3Nk$. If the electrons are free inside the lattice, they will contribute a further $\frac{3}{2}nk$ to the specific heat, for we have assumed that they form a classical gas of particles without internal degrees of freedom. But this contribution is not specially noticeable in practice; metals obey Dulong and Petit's Law (1819!) almost as well as insulators; the atomic heat tends to $3Nk = 3R$ at high temperatures.

This is a famous objection to the free-electron theory of metals. Indeed it is an insuperable obstacle within classical physics. If the electrons are free enough to carry electric currents then they must contribute those degrees of freedom to the specific heat.

A more subtle discrepancy having the same basic origin is in the *thermoelectric power*. According to classical theory, this should be much larger than is actually observed. We can deduce this quite simply from eqns. (1) and (3), where we expressed the electric current, and heat current, in terms of the drift velocity of the electrons. For any value of u, these have the ratio

$$\Pi = U/J \sim \tfrac{3}{2}\frac{kT}{e}. \tag{10}$$

Suppose now (fig. 4) that we establish a closed circuit of two metals, A and B,

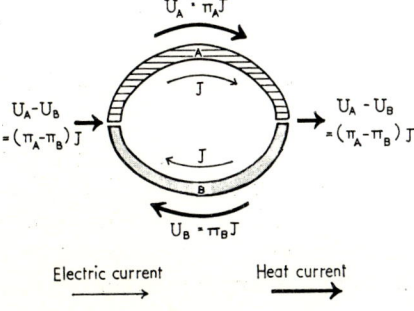

Fig. 4. The Peltier effect.

and make a current J flow round it. Even if we keep the whole circuit at nearly constant temperature, there will be a heat current drawn round with the electrons, of amount ΠJ. But suppose, for some reason, that Π is not the same in metal A as it is in B. Then the heat current $U_A = \Pi_A J$ arriving at the A—B junction will not be the same as $\Pi_B J$ leaving it. Ignoring Joule heat (which is quadratic in J, and can therefore be made negligible by making J small) we must reject the heat $(\Pi_A - \Pi_B)J$ at this junction. At the other junction the same amount of heat will be absorbed.

This is the *Peltier effect*, and (10) gives us at once the Peltier coefficient for our classical gas. There is a familiar thermodynamic argument (the Kelvin relation) which links this with the *Seebeck effect*, whose measure is the absolute thermoelectric power

$$Q = \Pi/T \sim \tfrac{3}{2}\frac{k}{e}. \tag{11}$$

Thus, in our free-electron model, all metals should have the same constant thermoelectric power, of magnitude about $100\,\mu\text{v}/\text{deg}$: the observed values are much smaller, up to, say, $10\,\mu\text{v}/\text{deg}$, and are proportional to the absolute temperature†. Moreover, in some metals Q is positive, instead of having the same negative sign as the charge on the electron.

7. The exclusion principle in a gas

The puzzle of the specific heat of the electron gas can only be solved within quantum theory. It is a consequence of the Pauli exclusion principle. No two electrons may be in the same state. In building up atoms we put one electron into each state (counting each orbital state as being doubled by spin); when we have filled up all the states in a shell we must move up to the next.

The question now is how to apply this principle to our free electron gas. How can we be sure that each electron is in a different 'state'?

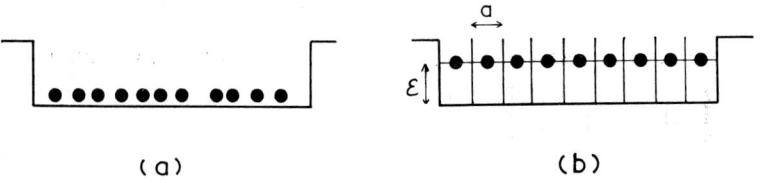

Fig. 5. (*a*) Classical jellium; (*b*) Cellular jellium.

One obvious way of doing this is to erect impenetrable walls about each electron, confining each to its own cell (fig. 5). But this localization costs energy. If the width of a cell is a, this is the maximum uncertainty in, say, the x coordinate of position of the electron. Correspondingly, the momentum of the particle in this direction will have a minimum uncertainty $\langle p_x \rangle$, such that

$$a\langle p_x \rangle \sim \pi\hbar. \tag{12}$$

---

† Of course this discrepancy could not have been apparent historically, as one could only measure *differences* of thermoelectric powers; the observed e.m.f.'s of thermocouples could then be written off as small differences from the classical formula for different metals.

Properly speaking, these uncertainties are root mean square deviations, so that we ought to write

$$\langle p^2_x \rangle \sim \frac{\pi^2 \hbar^2}{a^2}. \tag{13}$$

There will be similar rules for the y and z components. Adding these we obtain a formula for the mean total kinetic energy of the electron

$$\langle \mathscr{E} \rangle = \frac{1}{2m} \{\langle p_x^2 \rangle + \langle p_y^2 \rangle + \langle p_z^2 \rangle\} \sim \tfrac{3}{2} \frac{\hbar^2 \pi^2}{ma^2}. \tag{14}$$

(I have cooked the constants in the uncertainty principle to get the correct answer.)

This is a large energy—several electron volts. So we have a lot to gain, in our jellium model, by dropping the barriers and allowing the electrons to wander at will. Instead of keeping our states distinct in space, we allow them to overlap freely. In fact, we now go to the other extreme, and study electron wave functions covering the whole 'box' in which the gas is confined, the whole cube of metal. As in a typical atom, we must think about the 'different' wave functions which may be constructed in the same region of space, and then we put two electrons, one of each spin, into each such state.

This problem, of 'counting states' in a box, is an old one, which goes right back into the theory of elastic and optical vibrations. For our purpose the simplest argument is from the de Broglie formula: the wavelength of an electron of momentum p is

$$\lambda = \frac{2\pi\hbar}{p}. \tag{15}$$

Now suppose we have a cubical box of side La. We can get an exact number, s, of waves across the box if, say

$$s\lambda = La. \tag{16}$$

A wave of this wavelength, reflected from the side, and back again, would repeat itself exactly, and would thus be an acceptable stationary state of the system. If our wave is not travelling transversely across the box, there will be a similar rule governing the distance between peaks measured normal to the walls (figs. 6 and 7)

$$s\frac{\lambda}{\cos\theta} = La. \tag{17}$$

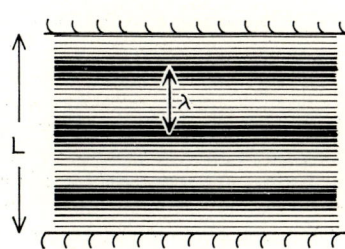

Fig. 6. Wavelength conditions for waves across a box.

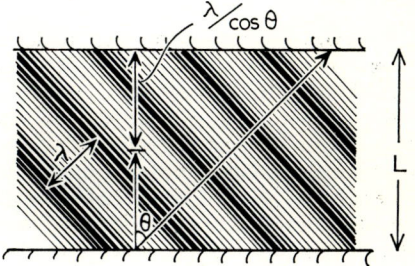

Fig. 7. Wavelength conditions for an arbitrary direction in a box.

Thus, if we take this normal as the x direction, we have a rule for the x component of momentum,

$$p_x = \frac{2\pi\hbar}{\lambda}\cos\theta = 2\pi\frac{\hbar s_x}{La} \qquad (18)$$

where s_x is an integer. There will be similar rules for the y and z components, each of which must be an integral multiple of $2\pi\hbar/La$. The states are 'quantized in momentum' with the quantum numbers s_x, s_y, s_z, which are +ve or −ve integers†.

We want to count states, and fill them in order. We must express the energy of each in terms of its quantum numbers s_x, s_y, s_z. Clearly

$$\mathscr{E}(s_x, s_y, s_z) = \frac{p^2}{2m} = \frac{4\pi^2\hbar^2}{2ma^2}\frac{(s_x^2 + s_y^2 + s_z^2)}{L^2}. \qquad (19)$$

In (12) the symbol a stood for the side of one little cubic cell containing one electron. If there are n electrons in the box of side La, then $n = L^3$. We could construct almost exactly n states by making each of the integers s_x, s_y, s_z run from $-\tfrac{1}{2}L$ to $\tfrac{1}{2}L$, for this would give us almost exactly L^3 different combinations of quantum numbers. In that case, the highest states would have the energy

$$\mathscr{E}(\pm\tfrac{1}{2}L, \pm\tfrac{1}{2}L, \pm\tfrac{1}{2}L) = \frac{3\pi^2\hbar^2}{2ma^2}. \qquad (20)$$

It is interesting that this is exactly the result that we derived in (14) for the energy of each electron combined to its own cell. This highest state is actually a solution to the problem with cell barriers, because these can be made its nodal surfaces. What we have done, in effect, is to replace 'localization in co-ordinate space' by 'localization in momentum space'. Because the electron is now allowed to range over a distance La we may define its momentum within the uncertainty limits $\pm \pi\hbar/La$. Whereas before, we kept each state distinct by not allowing the wave functions to overlap in real space, now we can only distinguish the different states if they differ in momentum by twice the limits of uncertainty in momentum. Instead of quantizing our system by a set of labels telling us which cell in real space the electron is in (i.e. the lth room in the mth corridor on the nth floor), we label each state by quantum numbers which tell us which cell it occupies in 'momentum space'.

This new quantization scheme does not change the energy scale. We cannot avoid a zero point energy of this order of magnitude by any means. But there is a net gain in freeing the electrons from their cells. The energy $\tfrac{3}{2}\pi^2\hbar^2/ma^2$ is the energy of *each* electron in the cellular scheme; it is the *highest*

---

† At this point let me say that I have cheated. If one is looking for stationary states in a box with perfectly reflecting walls these are *standing waves* like $\cos(xp_x/\hbar + \ldots)$, etc., which have nodes at every *half* wave length. For these p_x is quantized in units of $\pi\hbar/La$, so that we seem to have twice as many states as here. But then, of course, negative values of p_x, etc., only repeat the same range of functions, so must not be counted separately. It is much more convenient to split these into pairs of *travelling waves*, like $\exp(ixp_x/\hbar + \ldots)$ and $\exp(-ixp_x/\hbar \ldots)$, which have the same energy but are distinct mathematical functions. Travelling waves do not satisfy the 'reflecting' boundary condition, but one can set up other boundary conditions ('Born–von Kármán', or 'cyclic' conditions) which are not so intuitively obvious but which lead to the above quantization rules. There is a theorem that says that in a large box we get the same distribution of energy levels whatever the shape and condition of the boundary.

electron energy in the gas scheme. Thus, the *average* electron energy in the gas must be lower than this—in fact by $\frac{2}{5}\pi^2\hbar^2/ma^2$. This is quite a large energy—of the order of one or two electron volts in ordinary metals. Very crudely, we have here a source of the cohesive energy of metals. There are many complications in an exact calculation, but it is still approximately true to say that by bringing the atoms together and allowing the valence electrons to escape from their cells, we gain energy of about this amount. We can see that this cohesion depends, basically, on getting an electron gas of fairly high density, not on the details of the crystal structure. There is no need to construct special 'bonds' between neighbouring ions; the electron gas behaves as a ubiquitous liquid glue that will bind together more or less any arrangement of ions that is sufficiently densely packed.

This explains the tendency of metallic crystals to be 'close-packed', yet not strongly preferring one close-packed structure to another. We can understand the complex phase diagrams of alloy systems, where the balance of energy in favour of one structure may easily be tipped towards another phase by a small change of temperature or composition. The same general argument explains why metals are characteristically malleable and ductile. There are no specific bonds to be broken so that the atoms may be made to 'flow' plastically, without spoiling the cohesion of the solid.

Yet it must not be thought that quantization of momentum is necessarily the most favourable scheme for the valence electrons in all solids. In an ionic crystal, for example, the attraction of the spare electron from the metal atom to the halogen atom, and its localization there in a closed shell, easily outweighs the advantages of forming an electron gas. It is a nice point to decide the criteria favouring one or the other scheme, and to describe the transition between them that can be made to occur in some cases by compression or change of

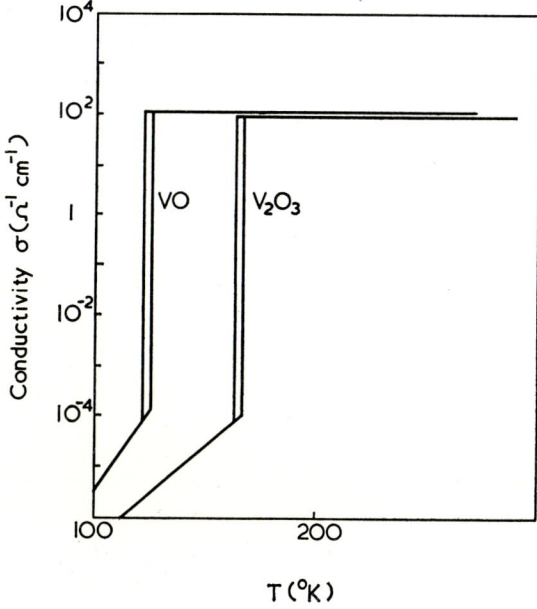

Fig. 8. The conductivity of oxides of vanadium, as a function of temperature (schematic).

temperature. The best evidence is that this transition is sharp. The valence electrons must either be closely localized, each to a region of the size of a unit cell of the crystal, or else they must be quite delocalized, in states that can only be specified by momentum variables. These two schemes are quite distinct, and there do not seem to be any intermediate cases.

8. The Fermi energy

The estimate (20) of the energy reached when we have put all the electrons into different momentum states is actually too high. In the first place, the distribution of quantum numbers $-\tfrac{1}{2}L \leqslant s_x,\ s_y,\ s_z \leqslant \tfrac{1}{2}L$ is not the optimum. We can see this by plotting out a point for each value of s_x, s_y, s_z (fig. 9), in a

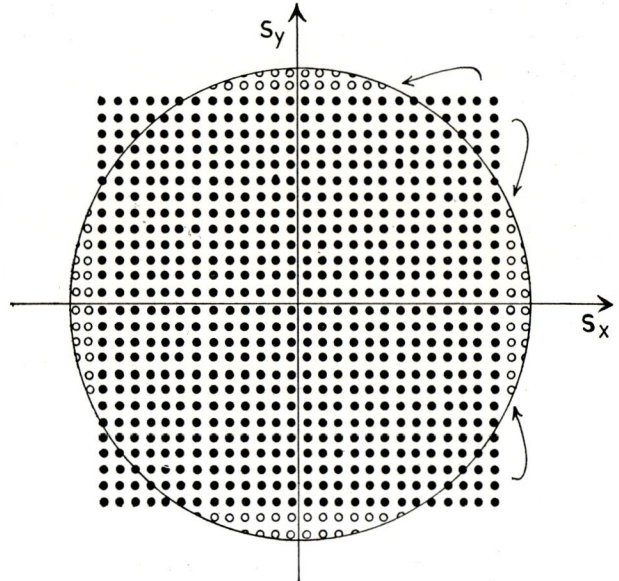

Fig. 9. The distribution of quantum numbers of allowed states in 'momentum' space.

three-dimensional space. These make an array of points, equally spaced, and the states in the range we have counted are those inside the cube (square) outlined in fig. 9. But from (19) we see that the energy of a state, in such a figure, is just the square of the distance of its representative point from the origin. But then points at the corners of the cube will have higher energy than those on the edges or faces. It is obviously favourable, energetically, to replace these higher states by others still empty and of lower energy. In fact, the best arrangement will be one where we only fill those states lying inside a sphere which contains the same number of points as our original cube.

Then again, each electron has two spin states and these are essentially distinct from the point of view of the Pauli principle. Thus, if we have n electrons per unit volume, we only need to use $\tfrac{1}{2}n$ distinct wave functions. So our sphere need contain only half the volume of our original cube. Ignoring slight errors due to the discreteness of our distribution of points, we can say that this sphere must have a radius D such that

$$\tfrac{4}{3}\pi D^3 = \tfrac{1}{2} L^3. \tag{21}$$

A point near the surface of the sphere would then have the energy

$$\frac{4\pi^2\hbar^2}{2ma^2}\left(\frac{D^2}{L^2}\right) = \left(\frac{3}{\pi}\right)^{\frac{2}{3}} \left(\frac{\pi^2\hbar^2}{2ma^2}\right). \tag{22}$$

This is the true energy of the highest occupied states in our distribution of electrons.

Let us write this energy in another form:

$$\mathscr{E}_\mathrm{F} = \frac{\pi^2\hbar^2}{2m}\left(\frac{3n}{\pi}\right)^{\frac{2}{3}} \tag{23}$$

where we have replaced a^3, the volume of a cubical cell that would contain one electron, by $1/n$. This shows us explicitly that the *Fermi energy* does not really depend on the size of the big cubical box or the size of the little cubical cells into which we divided the box. It is a function only of the density, n, of the electron gas, and not on such accidental features as the shape and size of the container. All the scaffolding of boxes, cells, stationary states and boundary conditions can now be removed, revealing this simple fact: if we apply the Pauli principle to a free-electron gas of density n, then the electrons must be supposed to fill states of kinetic energies up to $\mathscr{E}_\mathrm{F}$—an energy which is only a bit smaller than the energy each electron would have if it were confined to a cell of volume $1/n$.

9. Fermi–Dirac distribution

The Fermi energy is kinetic energy, of the order of several electron-volts, say 5 e.v. This is much more than the ordinary energy of thermal excitation: $kT = 0.03$ electron-volts at room temperature. In terms of electron velocity, the electron gas looks as if it is extremely hot, with a temperature

$$T_\mathrm{F} = \mathscr{E}_\mathrm{F}/k \sim 50\,000°. \tag{24}$$

The free electrons in a metal must be travelling with velocities about 1000 times the velocity of the ions, for their mass is smaller by a factor 10^4, and their energy larger by a factor 10^2. This is important for many theoretical investigations: we may usually treat the ions as if they were at rest, frozen into some typical configuration in the course of their thermal vibrations, whilst the electrons swarm about them. It is the basis of what is called the *adiabatic principle* in such calculations.

But although the electron system always has a high kinetic energy, it is not necessarily 'hot'. For example, it will still retain the high Fermi energy when it is at absolute zero, with the whole system in its ground state of energy. This is obviously just the state which we have assumed in calculating the Fermi energy: all the levels below this energy are filled and all the higher levels are empty. The system is obviously not 'hot', since it cannot give up any of its energy to a colder body.

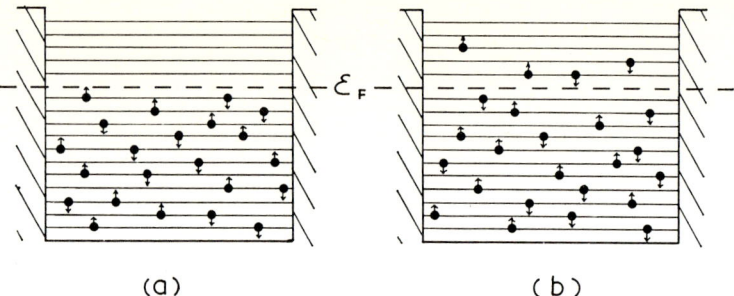

Fig. 10. Fermi–Dirac distributions (a) at $T=0$; (b) at a finite temperature.

Now if we heat this system, what can we do to the distribution of electrons? They can exchange energy with the lattice, in parcels of about kT at a time. This is easy enough for the electrons at the top of the distribution—they can move up into the empty levels above $\mathscr{E}_F$. For those deeper down, however, this is impossible. The levels above them are already full, and, by the exclusion principle, cannot take any more electrons. So these electrons are inhibited from exchanging energy with their surroundings, unless one can find a means of carrying them into an empty state above the Fermi level. This would require much more energy than one can get from ordinary thermal fluctuations. It would be, perhaps, an optical transition, of several electron volts.

Thus, although we have a very large number of electrons in our system, with kinetic energies ranging from zero to $\mathscr{E}_F$, when we heat it up we only affect those at the very top of the distribution. The same goes for any low energy excitation, such as application of an electric field, a magnetic field, or an acoustic vibration. The only electrons which can be affected are those in a narrow layer near the Fermi level. The other electrons can be treated as if they were quite inert, and no more capable of being excited than if they were in bound states deep down in the closed shells of the metallic ions. The information that we can obtain about the states of the electrons in a metal is thus limited. From the transport properties and some of the magnetic properties we can learn about electrons near the Fermi level; to study other states we must use optical excitation, or x-rays, whose effects tend to be brutal and difficult to interpret.

The formal theory of the statistical mechanics of a system such as this, of particles obeying the Pauli principle, is due to Fermi and Dirac. The rule is that the average occupation number of a state of energy $\mathscr{E}$ is given by

$$f(\mathscr{E}) = \frac{1}{\exp\{(\mathscr{E}-\mathscr{E}_F)/\boldsymbol{k}T\}+1}. \tag{25}$$

One can easily see that this number is almost exactly 1 if $\mathscr{E} \ll \mathscr{E}_F$, and that it is exponentially small (like the high-energy tail of the classical Boltzmann distribution) when $\mathscr{E} \gg \mathscr{E}_F$. The only region where $f(\mathscr{E})$ is appreciably different from 1 or 0 is near $\mathscr{E}_F$, over a width which depends on temperature, being a small multiple of $\boldsymbol{k}T$. We say that we have a *degenerate Fermi gas*, because $\mathscr{E}_F \gg \boldsymbol{k}T$. The names 'Fermi level', 'Fermi energy', and 'Fermi surface' come from this formula; Enrico Fermi himself did not make any other direct contributions to the theory of metals.

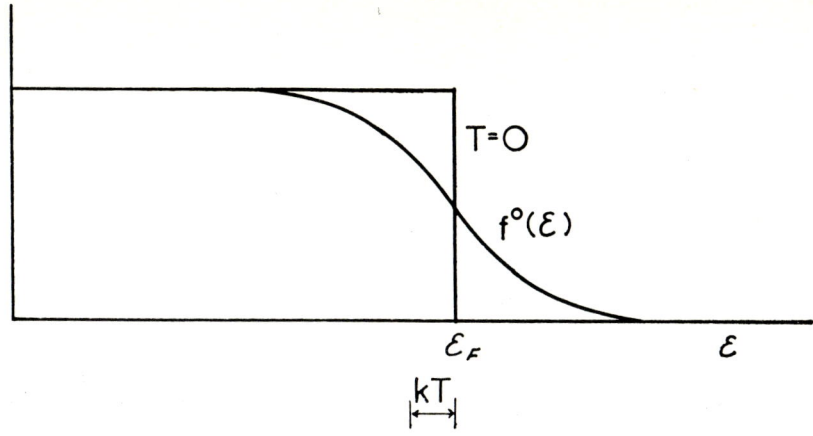

Fig. 11. The Fermi–Dirac function.

10. Electronic specific heat

We can now solve the puzzle of the specific heat of the electrons. When we raise the temperature to T, we do not increase the energy of *all* the electrons by the amount kT. Only a fraction of them, of the order of $kT/\mathscr{E}_F$, are affected. Thus, the extra heat energy per unit volume is

$$\delta W \sim \frac{kT}{\mathscr{E}_F} nkT. \tag{26}$$

This is equivalent to a specific heat of the form

$$C_{el} = \frac{\partial(\delta W)}{\partial T} \sim \frac{2nk^2 T}{\mathscr{E}_F} = \gamma T. \tag{27}$$

The electronic contribution to the specific heat is thus (a) not constant but linear in T and, (b) smaller by a factor like $kT/\mathscr{E}_F$ (e.g. 1/100) than the classical specific heat $(\tfrac{3}{2})nk$. It is not surprising that it cannot be noticed at ordinary temperatures. One must go to very low temperatures, where, according to the Debye T^3 law, the lattice specific heat becomes very small, before the linear term can be detected.

The smallness of the specific heat of the electrons also explains the small thermoelectric power of metals. In fact, if we introduce a factor $2kT/\mathscr{E}_F$ into (11), we find

$$Q \sim 3\frac{k}{e}\frac{kT}{\mathscr{E}_F} \tag{28}$$

which is a fair approximation, in order of magnitude and in dependence on temperature, to the observed thermoelectric power of a simple metal such as sodium. The electrons carry a much smaller net current of heat than we calculated in § 6, because their specific heat is so low.

Part II. Bands and Zones

1. Deficiencies of the free-electron model

A gas of free electrons in jellium is an excellent model for the electrical properties of metals. Indeed, it is so good, and so easy to use in practice, that extensive calculations have been made of all sorts of properties in addition to those which were discussed in I, and the same sort of semiquantitative agreement with experiment is found.

But really it is a bit bogus. How can we ignore the field of the ions—a field which is locally so strong that it would bind an electron in a free atom? Surely the conduction electrons must be scattered by the ions. How is it possible that this is not observed as a large constant electrical resistance? Then there are the experimental anomalies, such as positive Hall coefficients and positive thermo-electrical powers, which can by no means be explained within the free-electron model.

It is essential that we deal with the problem of the motion of an electron in the field of the lattice of ions. This seems a formidable task—which certainly it is if one wants to construct a programme that will print out values of the electrical conductivity and thermo-electric power of a metal, having fed into the computer only the potential field of each ion and the arrangement of the ions in the crystal lattice. It is only in the last few years that such direct quantitative calculations have produced results that are recognizably like the observed properties. But the formal structure of the theory has been well understood for the past 30 years, and it provides, in principle, answers to all the problems which I have just mentioned. In parts II and III, I shall try to explain this subtle and elegant theory, and to indicate the physical significance of its results.

2. Wave-propagation in a lattice

In our free-electron system, the electrons are represented quantum-mechanically by simple plane waves. Let us suppose that the atomic potentials are small, just to see their general effect on the wave functions. We might assume for example, that each ion is a small sphere, inside which the potential is constant, but a bit different from its value in the free space between the spheres. Each sphere will then scatter the electrons according to well known laws of diffraction theory.

But the spheres are arranged on a regular lattice—for the sake of argument, on a simple cubic array. The scattered waves from different atoms will tend to interfere with one another in a systematic way, so that there may be coherence in the total diffracted wave. This is precisely the same situation as for the

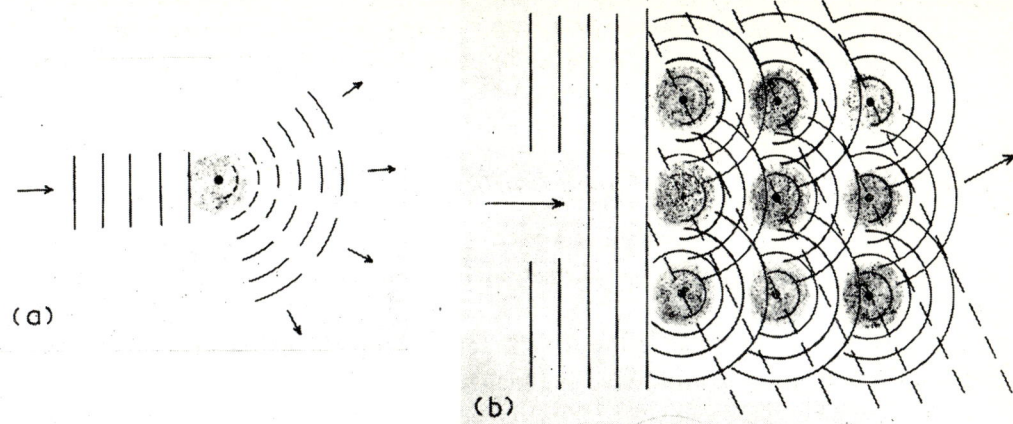

Fig. 12. (a) Scattering of electrons by a single atom.   (b) Diffraction by a regular array of atoms.

diffraction of x-rays, or fast electrons, by the crystal. It is well known that an x-ray beam will usually pass right through a crystal unless it has the wavelength corresponding to *Bragg reflection*, i.e. such that (see fig. 13)

$$n\lambda = 2a \cos \theta. \tag{1}$$

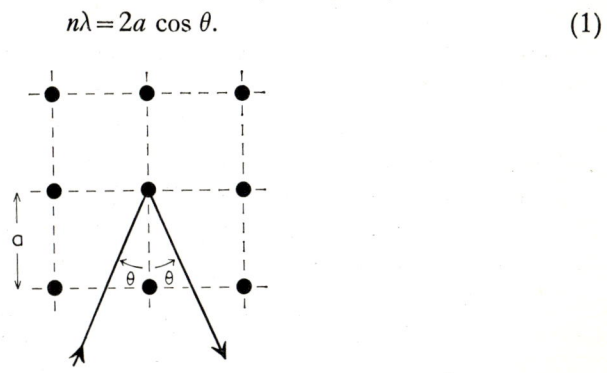

Fig. 13. Geometry of Bragg reflection.

For our electrons the very same law must hold. Using the de Broglie formula for the wavelength in terms of the momentum, we can write the condition in the form

$$pa \cos \theta = n\pi\hbar. \tag{2}$$

Any electron travelling in the direction θ relative to the lattice planes, with momentum p satisfying this condition, will be coherently reflected by the crystal lattice. For such an electron a simple travelling wave cannot possibly be a good representation of the wave function of a stationary state, for this electron will be continuously suffering internal reflections at the lattice planes.

3. Periodic potentials in one dimension

To simplify the discussion let us consider the case where the electron is travelling normal to a set of lattice planes. We then have, in effect, a one-dimensional problem, in which we are trying to propagate a wave through a linear array of objects, as in fig. 14. We need not concern ourselves in detail with the actual potential function for each such object; a set of square wells will do to simulate the obstacle presented by each lattice plane to the path of the

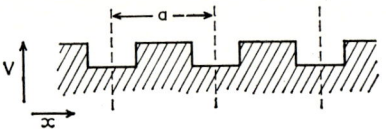

Fig. 14. Periodic potential in one dimension.

electron. The main point is that this potential repeats itself regularly in the x direction, with spacing a between the planes.

To eliminate the constant $\hbar$ from the algebra let us write $\mathbf{p} = \hbar \mathbf{k}$, so that $\mathbf{k}$ is the wave-vector, or, in our one-dimensional case, k is the wave-number, of the electron wave function. Thus, we start with a plane wave " travelling to the right ":

$$\psi_k = \exp(ikx). \tag{3}$$

If the electron is moving very slowly, k will be so small that the Bragg condition will not be satisfied. But for

$$k = \pi/a \tag{4}$$

the wave will be coherently reflected. This means that the wave function is transformed into

$$\psi_{-k} = \exp(-ikx), \tag{5}$$

which represents a wave travelling to the left. (The time factor in these waves is always assumed to be $\exp(-iEt/\hbar)$, but we need not include it explicitly in the formulae). We cannot avoid a mixture of such terms in the final wave function. Well then, consider uniform mixtures like

$$\psi_\pm = \frac{1}{\sqrt{2}} \{\exp(ikx) \pm \exp(-ikx)\} = \begin{pmatrix} \sqrt{2} \cos kx \\ \sqrt{2} \, i \sin kx \end{pmatrix}. \tag{6}$$

Since the scattering is symmetrical, to right or to left, ψ_+ and ψ_- will remain functionally unchanged by Bragg reflections. The effect of such scattering is simply to interchange the two travelling waves in the sum or difference. For this particular value of k, it looks as if the appropriate quantum–mechanical wave function for an electron in a steady state is a standing wave.

Now we could, of course, always construct such standing waves for *free* electrons by combining the two travelling waves as in (6). This is always permissible, because for free electrons both ψ_+ and ψ_- have the same energy. But in our periodic potential these two standing wave functions have *different* energies. We can prove this by actually calculating the average value of the potential energy of the electron in each state:

$$V_\pm = \frac{1}{L} \int |\psi_\pm|^2 V(x) \, dx$$
$$= \int \begin{pmatrix} 2 \cos^2 kx \\ 2 \sin^2 kx \end{pmatrix} V(x) dx. \tag{7}$$

The integration is over the length, L, of the ' crystal '.

It is obvious from the Bragg condition (4) that $|\psi|^2$ has the same periodicity as the potential function $V(x)$, so that these two integrals do not vanish. If we suppose that the mean value of $V(x)$ itself is zero (this is purely a matter of defining the energy zero of our system), then the two integrals will be equal in magnitude but opposite in sign. If, as in fig. 15(a), the peaks of electron density coincide

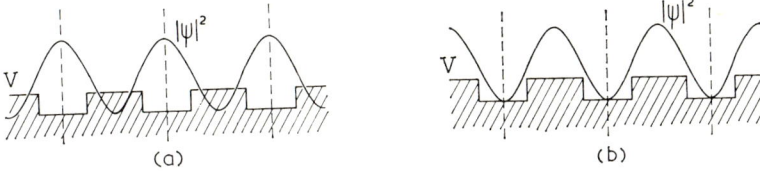

Fig. 15. (a) Wave-function with high electron-density in the potential wells. (b) Wave-function with high electron-density between 'atoms'.

with the centres of the potential wells the value of the integral will be negative; in the other case, where the electron density is zero at the centre of each well, the potential energy will be positive. In fact

$$V_{\pm} = \pm |V_g|, \tag{8}$$

where V_g is the Fourier component of the periodic potential at its fundamental wave number

$$g = 2\pi/a. \tag{9}$$

The two states ψ_+ and ψ_- will have the same kinetic energy as for free electrons:

$$\mathscr{E}^0{}_k = \hbar^2 k^2/2m. \tag{10}$$

Thus, the total energy of these two states is given by

$$\mathscr{E}_k(\pm) = \hbar^2 k^2/2m \mp |V_g|. \tag{11}$$

In other words, the state ψ_+ lies significantly lower in energy than a single free-electron state. It is therefore much more acceptable as a wave function for the electron†. But this state is a *standing wave*, so that the electron is effectively bound to the lattice, and is unable to travel as a wave packet throughout the crystal.

The result that we have just derived refers only to waves which exactly satisfy the Bragg condition—in other words, where $k = \pi/a = \tfrac{1}{2}g$. It is scarcely to be expected, however, that other waves will proceed through the lattice unmodified. The lattice is itself rather like a wave, of wavelength a, i.e. of wave number g. An electron wave tends to 'interfere' with this built-in potential wave, so that the state with wave-number k tends to be transformed into a wave of wave-number $k+g$ or $k-g$, etc. This is just what happens in Bragg reflection, where $k = \tfrac{1}{2}g$, and the reflected wave corresponds to $k+g = \tfrac{1}{2}g - g = -\tfrac{1}{2}g = -k$.

This suggests that the proper solution to the problem in the general case is not a travelling wave $\exp(ikx)$, but a linear combination of this with all possible 'reflected' waves i.e.

$$\psi_k = \alpha \exp(ikx) + \beta \exp\{i(k-g)x\} + \gamma \exp\{i(k+g)x\} + \dots \tag{12}$$

† Of course ψ_+ and ψ_- are not exact solutions to the Schrödinger equation in a potential such as fig. 15. One can easily construct such solutions in a series of square wells by joining together pieces of sinusoidal wave at each discontinuity of V. But (11) is a good approximation to the energy so long as $|V_g|$ is not too large, and the exact solution is still a standing wave.

where the coefficients are to be determined. This is a generalization of (6)—but we can no longer suppose that α and β have the same magnitude. This is because the 'reflected' wave, with momentum $(k-g)\hbar$, does not now have exactly the same kinetic energy as the original incident wave, so that it can only be present 'virtually' in the wave function. It must quickly be transformed, by a second reflection, back into the original wave, and will not therefore be present, on the average, with large amplitude.

As an approximation, let us take only the terms in k and $k-g$ in this sum, for these are the two which are most nearly equal in kinetic energy. We can easily calculate the total energy in such a state. The kinetic energy is just

$$\alpha^2 \frac{\hbar^2 k^2}{2m} + \beta^2 \frac{\hbar^2 (k-g)^2}{2m} = \alpha^2 \mathcal{E}^0_k + \beta^2 \mathcal{E}^0_{k-g} \tag{13}$$

since α^2 and β^2 represent the relative probabilities of finding the electron in these partial states. The potential energy comes out in the same way as in (7). Thus

$$\int |\psi_k|^2 V(x)dx = (\alpha^2 + \beta^2) \int V(x)dx + \alpha\beta \int \{(\exp(igx) + \exp(-igx))\} V(x)dx \tag{14}$$
$$= -2\alpha\beta |V_g|$$

because the average value of $V(x)$ is defined to be zero and the second integral is again just the principal Fourier component of the periodic potential.

We add together (13) and (14), and then choose the ratio α/β to make this total energy a minimum. The result is, by elementary algebra,

$$\mathcal{E}_k = \tfrac{1}{2}(\mathcal{E}^0_k + \mathcal{E}^0_{k-g}) - \tfrac{1}{2}\sqrt{\{(\mathcal{E}^0_k - \mathcal{E}^0_{k-g})^2 + 4|V_g|^2\}}. \tag{15}$$

This state has lower energy than the free wave and is therefore a more acceptable wave function for the electron in the lattice. Again it is not an exact formula but has all the important features of the exact solution.

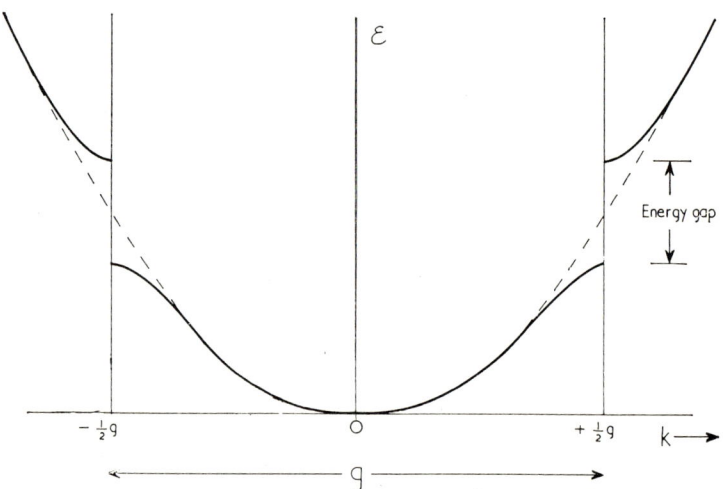

Fig. 16. Energy as a function of wave number for electrons in a one-dimensional periodic potential.

4. BANDS AND GAPS

To show the significance of this formula, let us plot $\mathscr{E}$ as a function of k. For reference, we first sketch in the free-electron parabola $\mathscr{E}^0{}_k = \hbar^2 k^2/2m$. Then we draw vertical lines at $k = \pm \tfrac{1}{2}g$, where Bragg reflections occur. The energy here is depressed below the free electron parabola by the amount $|V_g|$. If we evaluate (15) for other values of k, we find a function that starts almost exactly on the parabola at $k = 0$ but draws away from it smoothly until it reaches $\mathscr{E}^0{}_k - |V_g|$ at $\pm \tfrac{1}{2}g$—the points we have just plotted. If we were to look at the ratio β/α for different values of k, we should find that it was nearly zero near $k = 0$—i.e. the wave function is nearly a single travelling wave $\exp(ikx)$—and that this ratio increases to unity at $k = \tfrac{1}{2}g$—the reflected wave becomes more and more important until finally the travelling wave is brought to rest. We shall show later that this is no metaphor; as we approach Bragg reflection, the velocity of the electron falls to zero.

Nevertheless, there is no 'scattering' of the wave. If we start at the left hand end of the lattice with a wave which is the proper mixture of $\exp(ikx)$ and $\exp\{i(k-g)x\}$, but which corresponds, nevertheless, to a net flux of electrons to the right, then this wave is not *attenuated* by passing through any number of cells of the lattice. We shall show later that the *velocity* of propagation of wave packets along the lattice is not as high as it is for free electrons, and may even vanish for special values of k, but there is no question of the flux diminishing with distance as we go along the crystal, and no sign of a wave packet being reflected back, as it would be by an isolated square well. This gives us an answer to one of our basic difficulties: we have found that *there exist, in a regular periodic lattice, many solutions of the Schrödinger equation which are uniformly propagated without scattering or attenuation, and which can therefore carry an electric current without showing electrical resistance.* To that extent, the conduction electrons do not 'see' a perfectly regular lattice.

But now let us look at equation (15) from the point of view of energy, and ask what the electron state is for a particular value of $\mathscr{E}$. Near $\mathscr{E} = 0$, these are free-electron states. As the energy increases, the wave functions become more and more distorted, until, at $k = \tfrac{1}{2}g$, we have only a standing wave. Now what happens for higher values of $\mathscr{E}$? There is no solution of (15) in this neighbourhood. The fact is that, *for a whole range of energy there is no solution of the Schrödinger equation*. The next value above the level A at which a solution is allowed is at the level B, which corresponds, in fact, to the state ψ_- of (6). This is also a standing wave, now exactly out of phase with the lattice, and therefore of energy $+|V_g|$ above the free electron parabola. From here on, a continuous range of energy is allowed, corresponding to the function (15) with a positive sign before the square root. If we take these to be labelled by ever-increasing values of k, we soon find that we are back near the free-electron parabola, and, indeed back to nearly free-electron wave functions (fig. 16).

The effect of a Bragg reflection is thus perfectly simple. The relation between energy and momentum is no longer a continuous function. At the point where Bragg reflection would occur, the curve is split, with a jump discontinuity of magnitude $2|V_g|$. States of wave number less than $\tfrac{1}{2}g$ are lowered in energy; those of greater wave number are raised. An *energy gap* is created within which there are no solutions of the Schrödinger equation. If we look in detail at what happens when we try to construct such states, we find that 'k must become an imaginary number'. This is tantamount to saying that these solutions are real exponential functions, which will grow or decay so rapidly with distance that they cannot be normalized or stabilized except near the ends of our linear array of atoms. These are the 'Tamm states' or 'surface states'. They play a role in some semiconductor phenomena but are irrelevant to the properties of bulk metals.

This theory of wave propagation in a periodic one-dimensional array is not unique to metals. Readers may well be familiar with the 'pass bands' and 'stop bands' of frequency in the propagation of alternating currents through an array of electrical filters. The argument is precisely the same: for 'frequency', now read 'energy'. The elegant book by Brillouin† brings out the analogy to the full.

5. Brillouin zones

In § 2 we started out to study wave-propagation in a three-dimensional array of atoms, but found the problem a little too ambitious. We simply remarked that free electrons would suffer Bragg reflection if their de Broglie wavelength were equal to twice the projection of the spacing of a set of lattice planes on their propagation vector. But in § 3 and § 4 it was shown that the analogue of Bragg reflection in a one-dimensional lattice gave rise to a gap in the energy spectrum. Suppose now that we consider propagation in various directions, with various values of momentum, through a set of lattice planes—for example a set of planes spaced a apart, normal to the x axis (fig. 17). Let us use a wave vector $\mathbf{k} = \mathbf{p}/\hbar$ to specify the electron states, thus avoiding the word 'momentum' (which sounds too 'dynamical' in the present context) and also an extraneous factor $\hbar$. The condition for the 'first' Bragg reflection now reads

$$k_x = \pm \pi/a. \tag{16}$$

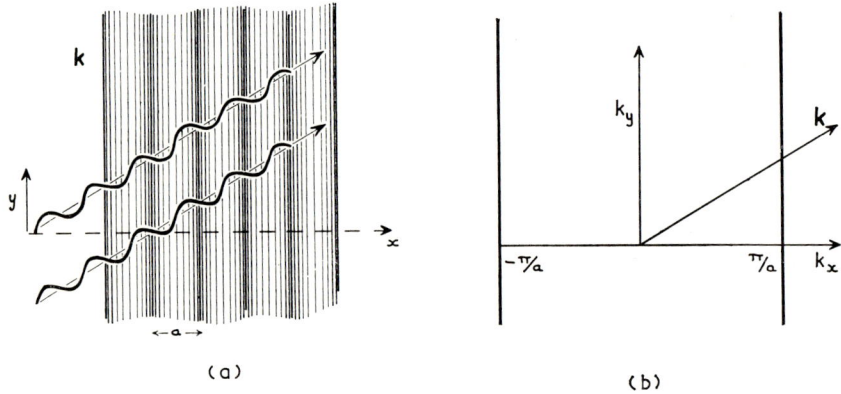

Fig. 17. (a) Propagation through a set of lattice planes. (b) Bragg reflection occurs when the **k**-vector lies on one of the planes $k_x = \pm \pi/a$.

But now in 'momentum space', or, what is now more appropriate, '**k**-space', this relation has a simple geometrical significance; the tip of the vector **k** lies on

† L. Brillouin *Wave Propagation in Periodic Structures* (New York: McGraw-Hill).

a plane normal to the x-axis, at a distance $+\pi/a$ or $-\pi/a$ from $\mathbf{k}=0$. An electron whose wave-vector reaches this plane will be Bragg reflected by the crystal lattice. In one dimension the analogue of these planes are the points $k = \pm \pi/a$ on the k-axis—and at these points we found an energy gap, bounded above and below by standing-wave states.

This suggests that the planes defined by (16) are *the locus of an energy discontinuity* in $\mathbf{k}$-space. That is to say, when we do, with great labour, set down and solve the Schrödinger equation for an electron in our lattice of ions, and find a function, $\mathscr{E}(\mathbf{k})$, which measures the energy in terms of the wave vector of the state, this function will be continuous from $\mathbf{k}=0$ up to these planes, and then will jump to a higher value. We can almost prove this directly by thinking that this set of lattice planes will scarcely affect wave propagation normal to the x-direction, so that we can choose our y- and z-components of $\mathbf{k}$, and write down the corresponding kinetic energy terms for plane waves, independently of k_x, and then we can think of the problem of constructing a suitable factor in the wave function for propagating in the x-direction just as if we were dealing with a one-dimensional system.

However, an ordinary simple cubic lattice has many different sets of lattice planes. The ones that naturally suggest themselves are the cube planes, spaced a apart along the x-, y- and z- axes. But now in $\mathbf{k}$-space there will be three sets

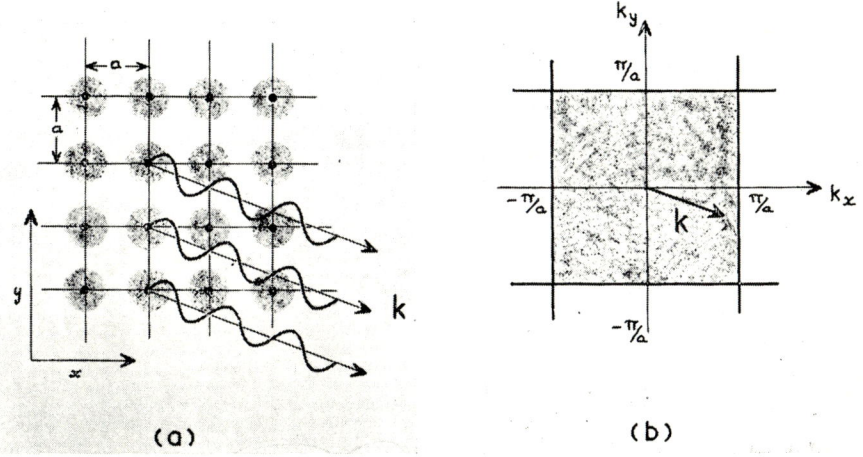

Fig. 18. (*a*) Propagation through a square array. (*b*) The Brillouin zone for a square array.

of discontinuity planes, forming a cubical box as in fig. 18. We call such a box a *Brillouin zone*, or simply *zone*. The planes of discontinuity of energy are the *zone boundaries*.

Here at last we are getting somewhere near the heart of the modern theory of metals. Let me repeat what we have done. We propose to study electron wave-functions in a periodic lattice. These are non-local wave functions spread through the whole crystal, so that each can have a well defined momentum or $\mathbf{k}$-vector. We therefore represent each state by a point in momentum space or '$\mathbf{k}$-space'. Because of coherent diffraction of the electrons by the regular atomic planes of the crystal, some states are very strongly affected by the periodic potential, and are very different from free-electron waves. The states which are most strongly diffracted, and whose energy is changed the most (up or down) are those whose **k**-vectors correspond to points on a set of planes which delineate a cell in **k**-space. This cell is a Brillouin zone.

6. Metal or Semiconductor

Brillouin zones have another property that is rather important. In I § 7 we considered the 'allowed momentum states' of a free-electron gas in a large box, so that we might count the number of states and find their energy. It was shown that the components of the momentum must be integral multiples of $2\pi\hbar/La$, where La is the length of the side of the big box representing the whole metal specimen. In terms of the wave vector

$$(k_x, k_y, k_z) = \left(S_x \frac{2\pi}{La}, S_y \frac{2\pi}{La}, S_z \frac{2\pi}{La} \right), \tag{17}$$

where S_x, S_y, S_z are integers.

But, of course L is a large number, so that this distribution of 'allowed states' in **k**-space is very fine-grained; for most practical purposes we take it as continuous. This set of rules will still apply when we treat a real metal, with a periodic lattice, in which the electrons are no longer quite free. Dynamical momentum itself is not a good quantum number, because of the interaction of the electrons with the lattice, but the vector **k**, which has many analogous properties, can be shown to be quantized according to the rule (17). (Some people refer to $\hbar\mathbf{k}$ as the *crystal momentum*). This only means that we have to fit integral numbers of wavelengths into the box, to satisfy the boundary conditions.

Now suppose that we have a simple cubic lattice, in which a is the lattice spacing. If we compare (16) and (17), we see that there is a close connection between the Brillouin zone boundary and the allowed states. The zone boundaries occur when the integers S_x, S_y, S_z have the values $\pm \frac{1}{2}L$. The zone is thus divided into a fine mesh of 'momentum cells', each corresponding to an allowed state (fig. 19). This is the same sort of picture as fig. 9 of I.

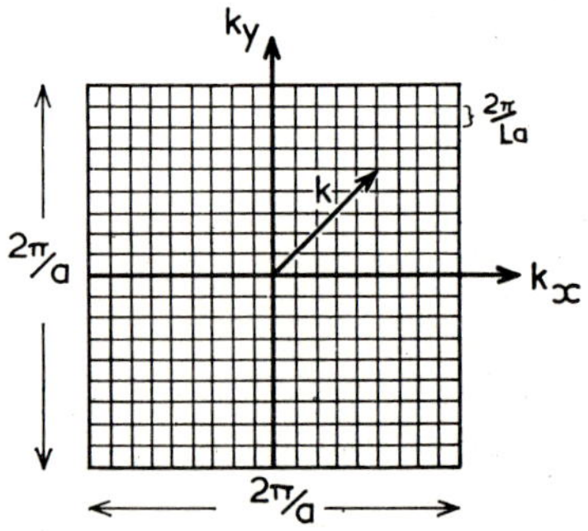

Fig. 19. 'Allowed' momentum states in a Brillouin zone.

The important point is that our cubical zone contains almost exactly L^3 such 'allowed states'. Between $-\frac{1}{2}L$ and $\frac{1}{2}L$, we have at our disposal L values of

S_x, L values of S_y, and L values of S_z†. In other words—*a Brillouin zone in* **k** *space contains as many 'allowed momentum states' as there are unit cells in the crystal lattice.*

It is important to grasp this clearly, as one may easily be muddled by the two different types of construction in **k**-space. The Brillouin zone is an invariant geometrical object. It is a polyhedron whose shape and dimensions depend only on the orientation and spacing of the lattice planes of the material. It is the same whatever the size and shape of the lump of material that is being studied. But the dimensions of the lump, and the boundary conditions on its surface, determine the fineness and detailed arrangement of the mesh-points of allowed states in **k**-space. This might be awkward; we might have to specify the size of our specimen in each experiment if it were not for two simple rules: the distribution of mesh points is always of uniform density in **k**-space, and the Brillouin zone always contains exactly as many points as there are unit cells in the specimen. From these rules we can enumerate states very simply. If our specimen contains N unit cells then we know that a fraction α of the volume of the Brillouin zone in **k**-space will contain αN allowed states of momentum. With two alternative states of spin, this volume could contain $2\alpha N$ electrons. In other words—a fraction α of the volume of a Brillouin zone can accommodate the electrons that would be in the solid if there were, on the average, 2α electrons per unit cell.

This suggests an interesting possibility. Suppose we have a divalent element which crystallizes into a lattice with one atom per unit cell. Then there will be exactly 2 electrons per unit cell—just enough to fill a whole zone right to the corners. This was the sort of arrangement we looked at in I, fig. 7—and found to be unstable. Electrons would 'flow' from the corners to the centres of the faces, until we had a sphere in **k**-space. But, if there were big energy gaps to overcome at these zone faces, this process would not seem possible, and an exactly full zone would be achieved. The electrons in this zone would fall into a full energy band, with a gap above it.

A full band is electronically inert—rather like a closed shell of electrons in an atom. The element would not be a metal, but a *semiconductor*. Any electrical conductivity could be traced to minor deviations from the perfect closed-band state. A few extra electrons may be introduced along with chemical impurities, or there may be thermal excitation of a few electrons above the energy gap into the next band, where they are more or less free to carry electrical currents. By comparison with a metal, the electrical conductivity would be small, and very sensitive to changes of temperature and to traces of dirt. If the energy gap is very large (diamond is an example), the solid may even be effectively an insulator. In such a case we might begin to wonder whether localized states for the electrons, as discussed in I § 7, might not be more appropriate. The 'energy gap' may then be thought of as an 'ionization energy' or 'electron affinity' for the release of an electron from the atom or ion at an individual lattice site.

As it happens, all Group II elements are metals: we must go to Group IV of the periodic table, to carbon, silicon and germanium, to find a semiconductor. The argument that I have just presented assumes that the bands do not overlap.

† Strictly speaking, $(L+1)$, if we include the two end points; the difference is insignificant because L is very large.

This is the case in one dimension. But in three dimensions we have to study the whole zone. Although there will be an energy gap when we cross the zone boundary at any point, this gap does not come at the same absolute energy at all points on the boundary. It may be (fig. 20) that the top of the gap at some point A lies actually below the bottom of the gap at some point B. Then, as we add electrons, they will 'spill over' into the upper band of states around A, instead of filling right up into the corner at B. This means that the lower band is not

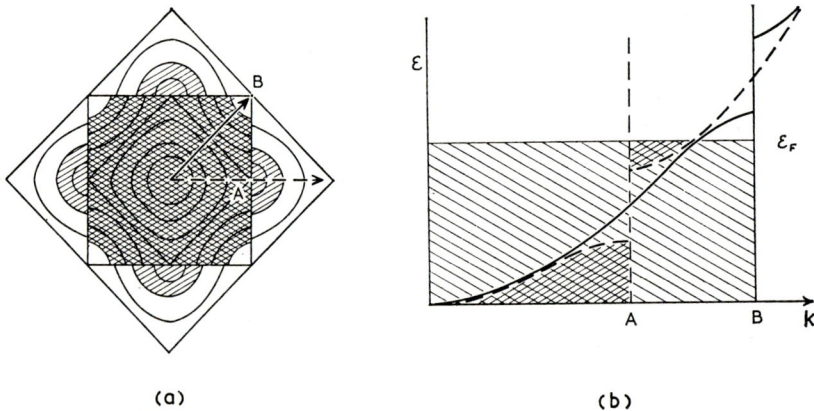

Fig. 20. (*a*) Energy contours, and a Fermi surface for a divalent metal. (*b*) Cross-sections of the Brillouin zone along two directions, showing overlap of bands.

quite full, and the upper band not quite empty. The electrons in these bands are no longer energetically isolated under energy gaps, and can easily be made to carry electric currents etc.

7. The Fermi surface

The essence of the argument of this chapter is that the periodic lattice has the effect of carving the energy distribution of the conduction electrons into a series of bands. Each band can contain two electrons per unit cell. From this alone we can argue that an element such as Cu, with one electron per atom, one atom per unit cell, must be a metal, for there simply are not enough electrons to fill a band. The same goes for Al, with the same crystal structure and three electrons per atom. But Mg with two electrons per atom, and two electrons per unit cell, could be a semiconductor—if it were not for large overlaps across the zone boundaries. Some of the tetravalent elements are metals, some are semiconductors, because in some cases the bands overlap whilst in others they are separated by a substantial energy gap. The case of As, Sb and Bi is interesting. With 5 electrons per atom, we might anticipate simple metals. But they crystallize in a structure with two atoms per unit cell, so there are really 10 electrons to accommodate. As it happens, the 5 bands are nearly all full, but there is a little overlap (in Bi, about 1/1000 of an electron per atom!) into a 6th band. These elements are such poor conductors that they are called 'semi-metals.'

It is also interesting to notice what happens when we melt a solid. The crystal lattice disappears and the sharp Bragg reflections are smudged out into one or two broad peaks. The Brillouin zone boundaries lose their meaning, and the energy gaps more or less disappear. Thus, the bands coalesce, and we have an electron distribution which is effectively the same as in a free-electron gas. We find that semi-metals and semiconductors become metallic when they melt, so that the liquid Pb and liquid Ge have nearly the same electrical properties.

One can evidently give a good qualitative account of the properties of metals from an analysis of their band structure. One can even go further, and construct curves representing the distribution of electrons in energy in different bands—the function being the *density of states*, defined such that $\mathcal{N}(\mathscr{E})d\mathscr{E}$ is the number of electrons states in the range of energy $\mathscr{E}$ to $\mathscr{E}+d\mathscr{E}$ (fig. 21). If $\mathcal{N}(\mathscr{E})$ is small at the Fermi level, then we deduce that the metal is not a good conductor,

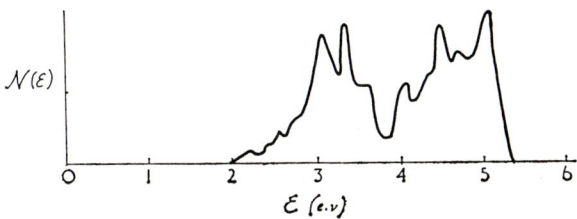

Fig. 21. A theoretical estimate of the density of states in the *d*-band of a face-centred-cubic metal.

and so on. But for a *quantitative* account the band picture is inadequate. It does not allow for the differences between the properties of different electron states at the Fermi level itself. In fig. 20, for example, the states near the corners of the zone are entirely different from those that have been filled by overlap in the second band. In more complicated cases, electrons that are nominally in the same band may not be equivalent, and must be treated separately in calculating the macroscopic properties.

But we are still only interested in states near the Fermi level, and these all lie on or near a surface of constant energy in **k**-space. This surface, defined by the relation

$$\mathscr{E}(\mathbf{k}) = \mathscr{E}_\mathrm{F}, \qquad (18)$$

automatically provides for us all the electron states that can play any part in the ordinary transport properties of the metal, or in such thermal properties as the specific heat. It is called the *Fermi Surface*. The name has been traced back to a paper by Bardeen, in 1940, but the concept can be found in much earlier papers, as long ago as 1930. From a mathematical point of view, **k**-space is the natural medium for the classification of electronic states, and the construction of surfaces of constant energy is then irresistible. We automatically talk of 'Electrons on the Fermi surface', instead of 'electrons at the Fermi level', and when we cool our metal to absolute zero we say that 'all states inside the Fermi surface are full; all those outside it are empty'.

In a free electron gas, of course, the Fermi surface is a sphere. This is essentially the construction of I, fig. 9. The geometry of a sphere is so simple that we need scarcely use a concrete model of this sort, and can calculate the electronic properties explicitly by analytical methods†. But in a real metal the Fermi surface is almost sure to be different from a sphere. Even in a monovalent metal the energy in certain directions in **k**-space may be so much reduced by the energy gaps at the nearest zone boundaries that the surface is drawn into contact with the zone faces. If there are two or more electrons per atom, the influence of the zone boundaries is sure to be profound, for they will certainly intersect any free-electron sphere that we draw, and thus will introduce energy gaps at every line of intersection.

What this means is that the properties of the electron gas will not be isotropic. The dynamical properties of an electron at the Fermi level will depend on its direction of motion. If it is travelling in a direction where it might very easily suffer Bragg reflection by the lattice, then it may seem 'slow', and 'heavy'; in other directions it may behave very like a free electron. These dynamical properties just depend on ' where it is on the Fermi surface ', so that in many cases the shape of the Fermi surface relative to the Brillouin zone is a guide to the electrical properties of the metal.

This is obvious in the pre-war work. What is new, and exciting in this field is that the Fermi surface can be treated as if it were almost a solid object. It is as invariant as the Brillouin zone within which it is defined. By various elegant experiments we can trace out its shape with almost the same accuracy as we can survey a range of mountains. It is difficult to think of a case in the history of physics which better exemplifies the principle that ' last year's abstract mathematical construction is this year's concrete model of reality '.

---

† Because of this analytical simplicity, the free electron model lasted too long. Even when it was breaking down, attempts were made to save the algebra by ' generalization ' to ellipsoidal surfaces. It is now obvious that Fermi surfaces are usually much more complicated than this—may, for example, be multiply-connected.

Part III. Dynamics of Bloch Electrons, and the Calculation of Band Structure

*' Not in utter nakedness, nor in sheer forgetfulness
But trailing clouds of glory do we come '*
—Wordsworth.

1. The need for 'Theory'

In this part we shall consider two problems that were raised in I, and not resolved in II: how is it possible that sometimes the carriers in a metal behave as if they were positively charged, and how do we set about actually calculating the shape on the Fermi surface in practice? The first question is easily answered by a piece of formal analysis; the interaction with the lattice can have a decisive effect on the dynamical properties of an electron. The second question requires an exposition of the technique of what is usually called ' The calculation of band structure '. It may seem that this is a somewhat specialized mathematical subject, of interest only to the professional theoretical physicist. So it seemed until recently. But the latest results in this field are of great value in giving us a physical insight into ' what goes on ' in a metal. We can now, I think, see through the elaborate mathematical formulae, and understand instinctively why the electrons are not scattered much more strongly by the ions in a metal lattice, and why interaction between the electrons is also, in practice, relatively unimportant.

It is also worth remarking that a good deal of such theory preceded the discovery of experimental methods for studying the Fermi surface. The interpretation of the experiments is not easy, and the guidance of theory is essential. The most fruitful work in this field (as in most branches of physics) has come from close co-operation between calculators and observers, between theoretical and empirical science.

2. Dynamical properties of electrons

For the calculation of electrical conductivity, we need to know the *velocity* of the electrons on the Fermi surface. In a free-electron gas this presents no problem; the velocity is proportional to the momentum:

$$\mathbf{v} = \frac{1}{m}\mathbf{p} = \frac{\hbar}{m}\mathbf{k}. \tag{1}$$

The effect of Bragg reflections is to hinder the linear motion of the particle, so that, on the average, its velocity may be much less than we should calculate from (1). There is a very simple formula for this effect. In wave mechanics the wave function is not constant, but has a time-dependent factor. An electron in the state of energy $\mathscr{E}(\mathbf{k})$ is supposed to have a ' frequency '

$$\nu(\mathbf{k}) = \mathscr{E}(\mathbf{k})/\hbar. \tag{2}$$

This 'state' is thus a travelling wave system extending through the whole crystal. But to discuss the 'velocity' we must partially localize the electron, and then watch how it moves. To do this we need to construct a wave-packet, by superposing wave functions from different states in the neighbourhood of **k** (i.e. of slightly different wavelengths) in phase at our point of localization, and then we watch how this wave packet moves.

We now have a familiar problem of diffraction theory. The formula (2) is essentially a dispersion law, relating the frequency of a wave to its wave number k. It is well known that the *group velocity* in such a medium is

$$v = \frac{\partial \nu(k)}{\partial k} = \frac{1}{\hbar} \frac{\partial \mathscr{E}(k)}{\partial k}. \tag{3}$$

At least, this is the isotropic case, which is typical of most acoustic, hydrodynamic and optical media. More generally, however, we have to think of the possibility of somewhat anisotropic media, where the group velocity is not the same, for a given wave packet, in all three coordinate directions. In that case we treat each direction separately, and write

$$(v_x, v_y, v_z) = \frac{1}{\hbar} \left(\frac{\partial \mathscr{E}(\mathbf{k})}{\partial k_x}, \frac{\partial \mathscr{E}(\mathbf{k})}{\partial k_y}, \frac{\partial \mathscr{E}(\mathbf{k})}{\partial k_z} \right), \tag{4}$$

defining each component of velocity as equal to the derivative of the energy with respect to the corresponding component of the wave vector. In other words we treat $\mathscr{E}(\mathbf{k})$ as a continuous function of position in **k**-space. The velocity of an electron 'in the state **k**' is then simply the local gradient of $\mathscr{E}(\mathbf{k})$ in that space:

$$\mathbf{v}(\mathbf{k}) = \frac{1}{\hbar} \operatorname{grad}_\mathbf{k} \mathscr{E}(\mathbf{k}). \tag{5}$$

This result is of great importance and interest. Once we have, by some means, calculated or discovered the form of $\mathscr{E}(\mathbf{k})$, the simple operation of taking local derivatives gives the electron velocity. The whole effect of the crystal lattice is contained in the function $\mathscr{E}(\mathbf{k})$; there is no need to go back and look at

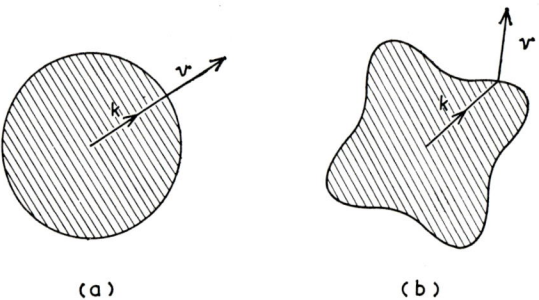

(a) (b)

Fig. 22. For a non-spherical Fermi surface, the velocity of an electron may not always be parallel to its propagation vector.

the actual form of the wave function. Moreover, the velocity of an electron is always normal to the energy surface on which it lies. From the shape of the Fermi surface we can calculate this direction at any point. For free electrons, where the Fermi surface is a sphere, it is obvious that the velocity is always parallel to the momentum. For 'Bloch electrons'—conduction electrons in metals, where the propagation is seriously affected by the crystal lattice—this surface may be seriously distorted, and **v** is not at all sure to be parallel to **k**.

To calculate the *acceleration* of an electron (i.e. of a wave-packet such as we have considered) we need a more formidable analysis. The result may be quoted: a force **F**, such as an electric field, changes the state of the electron in such a way that

$$\hbar \dot{\mathbf{k}} = \mathbf{F}. \qquad (6)$$

This is a Newtonian law—the force equals the rate of change of crystal momentum. It sounds plausible, so long as we are quite sure that $\hbar \mathbf{k}$ really is the analogue of momentum for an electron in a Bloch state, but a good deal of hard work has to be done to justify it exactly. It is now believed to be true, except in an external field so enormous as to rival the local field around each metal ion.

In classical electrodynamics, the force on a moving electron in electric and magnetic fields is given by

$$\mathbf{F} = e\left(\mathbf{E} + \frac{1}{c}\mathbf{v}_\wedge \mathbf{H}\right). \qquad (7)$$

Again, we assume this *Lorentz force* formula for our Bloch electron, and write

$$\hbar \dot{\mathbf{k}} = e\mathbf{E} + \frac{e}{c}\mathbf{v}(\mathbf{k})_\wedge \mathbf{H}. \qquad (8)$$

The proof of this formula in a periodic potential is a major problem of quantum theory, only recently solved. Fortunately there is overwhelming experimental evidence that it is very exactly true in all ordinary fields.

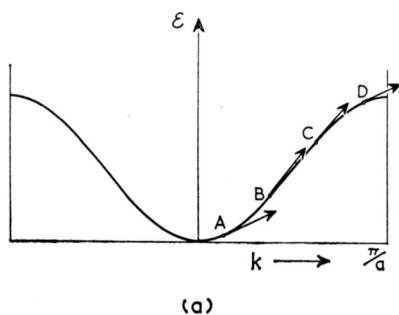

Fig. 23. (a) Energy in a one-dimensional crystal.

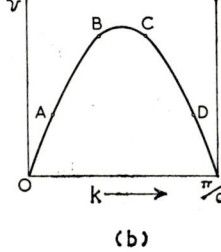

(b) Electron velocity in a one-dimensional crystal.

3. 'ELECTRONS' AND 'HOLES'

The formulae of the previous section have a rather peculiar consequence. Consider a one-dimensional crystal, for which $\mathscr{E}(k)$, in a single band, takes the form shown in fig. 23. Now apply an electric field such as to accelerate a free electron to the right. For an electron near the bottom of the band everything is quite simple. The value of k increases with time, the electron moves from a state such as A to a state such as B, and its velocity increases. We could write down a formula for the actual 'acceleration' i.e. rate of change of velocity—by combining (3) and (8)

$$\dot{v} = \frac{\partial}{\partial t}\left(\frac{1}{\hbar}\frac{\partial \mathscr{E}(k)}{\partial k}\right) = \frac{\partial k}{\partial t}\frac{\partial}{\partial k}\left(\frac{1}{\hbar}\frac{\partial \mathscr{E}(k)}{\partial k}\right) = eE\frac{1}{\hbar^2}\frac{\partial^2 \mathscr{E}(k)}{\partial k^2}. \tag{9}$$

It is as if the electron had the mass m^*, where

$$\frac{1}{m^*} = \frac{1}{\hbar^2}\frac{\partial^2 \mathscr{E}}{\partial k^2}. \tag{10}$$

But near the top of the band the situation is quite different. An electron 'accelerated' from C to D, say, finishes up with a smaller velocity than before the electric field was applied. The increase of k has brought it nearer to the zone boundary, nearer to Bragg reflection, and has thus actually slowed it down. The formulae (9) and (10) are still formally true, but the curvature of $\mathscr{E}(k)$ is now negative so that *the 'effective mass', m^*, would be a negative quantity.*

From a mathematical point of view there is no difficulty about this, but it is inconvenient when we try to apply to such states our intuitive notions and experience of the dynamics of ordinary particles. The following trick† saves us. Suppose we think about the properties of an *empty* state in an otherwise full band. Such a state will behave as if it had *minus* the mass of the electron that would have filled it. If the curvature of $\mathscr{E}(k)$ is negative, then this 'hole' in the full band will behave like a normal Newtonian particle of positive mass. But it will also have *minus* the charge of the electron that would have filled it, so it will behave electrically as a *positively* charged particle.

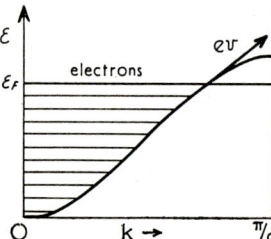
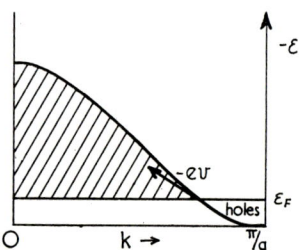

Fig. 24. A band that is nearly full of 'electrons' may be described as a band containing a few 'holes'.

This analysis is so like the now-familiar Dirac theory of the positron that we need scarcely pause to think it out in detail. To be consistent we should measure 'the energy of a hole' negatively, that is, downwards from the top of the band. We have replaced a 'band nearly full of electrons' by a 'band containing a few

---

† It is more than a trick: the most compact wave-packets at the top of a band are actually these 'hole' states.

'holes', with a Fermi level, velocity, etc. clustered around the edge of the zone. We can think of a 'hole' as an ionized atom—one that has lost an electron—in a solid where otherwise the sites are occupied by simple neutral atoms. There is a tendency for an electron from a neighbouring atom to fall into the ion, leaving the original site ionized. Thus the 'state of ionization' can move through the lattice like a positive charge. This motion can have its own kinetic energy, and thus behaves like a dynamical particle. Where there are only a few holes in a band, this description is much more economical than one where we discuss the motion of all electrons into and out of the various sites.

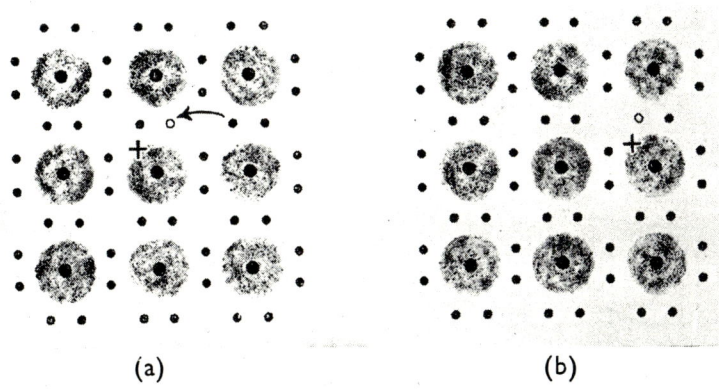

Fig. 25. The motion of an electron to fill the vacant orbital is equivalent to the motion of a 'hole' carrying a positive charge.

We can now understand some of the 'anomalous' properties of metals which could not be explained inside the free-electron model. Consider, for example, the Hall effect. As we saw in I, § 5, the direction of the Hall field depends on the sign of the charge carriers. Positively-charged holes can thus be detected by their Hall field being opposite to that of ordinary electrons. This effect is well known in semiconductors, where the holes at the top of the valence band are very important as carriers in 'p-type' material.

But we can see the same effect in a divalent metal, where the zone is nearly full except for regions near the corners and overlaps through the zone faces (see fig. 20). The bits of Fermi surface in the overlaps will be electron-like, because $\mathscr{E}(\mathbf{k})$ is near a local minimum, and thus has positive curvature in that neighbourhood. The bits in the corners are best described in terms of holes, for $\mathscr{E}(\mathbf{k})$ has a maximum in each corner, and thus has negative curvature (i.e. negative effective mass). There are as many holes as electrons, so that such a metal would have negative or positive Hall coefficient according as electrons or holes contributed more to the electric current. To calculate this in detail is difficult, but we can certainly now understand in principle why the Hall constant is positive in Be, Zn and Cd, and small in Sn and Pb.

The same argument applies, in a general way, to the thermoelectric effect. A current of holes will tend to give a positive thermoelectric power, as observed in some metals. But here the situation is really much more complicated, and it does not follow that a positive thermoelectric power in a metal is a sign of the predominance of 'holes' in the electrical conductivity. We shall return briefly to this point in IV.

It must not be thought that the separation into electrons and holes is unique and unambiguous. It is perfectly possible for $\mathscr{E}(\mathbf{k})$ to have "saddle points" where it is stationary but is neither a maximum nor a minimum. In such regions the curvature of the surface may be positive in one direction and negative in

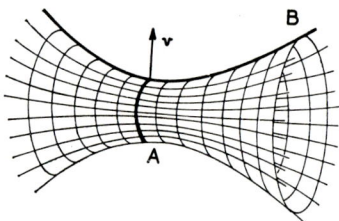

Fig. 26. The state with velocity **v** will behave like an 'electron' in the plane A, but like a 'hole' in the plane of B.

another. The dynamics of states in this region may be most unexpected. It is a nice problem to calculate the laws of scattering of such an object in a central field of force!

4. The problem of calculating the band structure

So much for the in-principle problem of how it can be possible for apparently simple metals to show such anomalous behaviour as a positive Hall coefficient. But now we want to make a detailed quantitative study of each metal, and to derive numbers for its observed electrical properties. We want to know its band structure, or, for greater accuracy, the function $\mathscr{E}(\mathbf{k})$ which fixes the shape of its Fermi surface, electron velocities, etc. As we shall see in IV and V, there are empirical methods for determining the shape of a Fermi surface in detail, but we shall never be satisfied until we can also derive this shape by direct calculation from the known crystal structure and known potential of each atom or ion. In practice it is also almost essential to have some idea of the shape of Fermi surface from theory before one can disentangle the strong threads of evidence derived from such phenomena as the de Haas-van Alphen effect and the high-field magneto-resistance.

In our one-dimensional model, we found a solution very simply, by considering the effect of Bragg reflection from the regular lattice array. The result was expressed in terms of one Fourier coefficient of the lattice potential. But this was only an approximation, in which it was assumed that we could write the wave function in the form

$$\psi_k = \alpha \exp(ikx) + \beta \exp[i(k-g)]x. \tag{11}$$

In fact the wave function must be much more complicated than this. Because the ionic potential is very strong near the nucleus, the electron must have a high kinetic energy there, to avoid being captured. It must, indeed, look very like the wave function of a valence electron in the free atom, with, say, 2 or 3 nodes inside the closed shells of the ion. A formula such as (11), combining just two waves of wavelength greater than the lattice spacing, is quite inadequate to represent such rapid oscillations. Our primitive solution cannot be generalized to provide a reasonable means of solving the practical problem. It is only valid when the effect of the lattice is a small perturbation; the actual field of the ions is too strong.

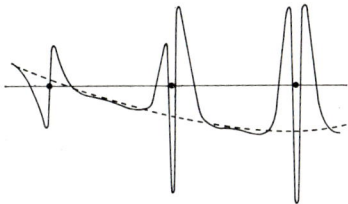

Fig. 27. Typical wave-function of an electron in a crystal.

5. Tight-binding and cellular methods

Well then, let us exploit our expectation that inside each ion the wave function must be rather similar to the wave function of an electron in the free atom. We can, for example, construct a whole set of these functions, $3s$, $3p$, $3d$, $4s$, $4p$ .... etc., around each ion in the lattice. Then we can take linear combinations of these, with appropriately wavy factors, and look to see how they satisfy the Schrödinger equation in the ionic potential. From a formal point of view, this is quite easy, and physically speaking it is the wave-mechanical analogue of our original argument for free electrons. We are simply setting up the problem on the assumption that an electron 'hops' from ion to ion, and is eventually shared out amongst all the ions of the crystal.

This method is called the *tight-binding method* because it starts with each electron tightly bound in its own atom, and then allows the bonds to be loosened by interaction with neighbours. We can see the effect in a schematic way if we

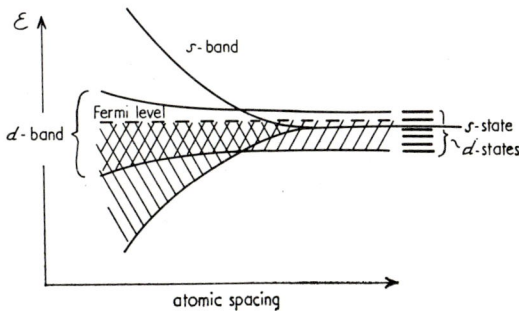

Fig. 28. States of the free atom combining to form bands.

take an array of atoms spaced far apart, and gradually bring them together so that the interaction between neighbours increases. We find (fig. 28) that each discrete level of the free atom tends to give rise to a whole band of states, the width of the band broadening as the atoms come closer and interact more strongly. The 3s-state in each of N atoms is two-fold degenerate because of spin. When the atoms are brought together, these $2N$ states combine and spread apart in energy to make a band, which we naturally call the 3s-band. Thus, we can identify the various energy bands in a metal by the atomic levels from which they arose.

This principle is exemplified in the transition metals. The electrons in the d-levels combine to form a 'd-band' which is more or less distinct in properties from the band formed by the outer s-electrons (although they may overlap in energy). If this d-band is not full it will contribute to the electrical properties of the metal, but in quite a different way from the s-band, where the electrons behave as if nearly free. The d-wave functions are drawn rather compactly about each atom, so that they do not interact strongly with their partners on neighbouring atoms. The d-band is thus rather narrow, and yet may contain a high density of electrons. The velocity of the d-electrons must be very small, so that they do not carry much electric current, and so on. But here we enter a controversial field. The electrical structure and properties of the transition metals are not yet understood.

Unfortunately the tight-binding method is not satisfactory as a practical means of calculating the band structure. The wave function is well represented *inside* the ion core (i.e. inside the closed shells), but the method cannot deal with the large proportion of the volume of the metal *between* the ions. In these interstitial regions the potential is nearly constant, the kinetic energy of the electron relatively small (e.g. the Fermi energy) and the wave function rather smooth. Atomic orbital wave functions die off exponentially outside their atoms; this is too drastic.

The situation can be saved to some extent by constructing *cellular* wave functions. We draw a unit cell around each atom, and then solve the Schrödinger equation under new conditions—for example that the wave function in each cell should join smoothly on to the wave-function in the neighbouring cell—in practice, on to itself again with a phase factor. These boundary conditions can be made explicit, and the problem can be set up and solved systematically. The results are quite good in practice, but the problem tends to be ' over-determined ' —the conditions for smoothness of join across the whole plane separating two neighbouring cells are an infinite set, so that it is difficult to choose from them a small number of the most significant relations to determine the coefficients in the linear combinations of solutions. The computational difficulty emphasizes the weakness of the method in principle. We have introduced an artificial boundary in the region where the potential is actually most nearly constant.

6. Muffin-tin potentials, APW's and OPW's

We can come closer to the real physical situation if, instead of drawing artificial boundaries between the cells of the crystal we deliberately treat the region between the ions as empty space in which the electrons propagate as simple wave planes. Around each ion we draw a sphere, as small as we dare, inside which the potential is spherically symmetrical and rather like the potential in a free atom. In the interstitial region the potential is constant. The potential map then looks like a cake-, cookie-, or muffin-tin (according to one's continent), with each circle-pattern the site of an ion.

Fig. 29. Muffin-tin potentials.

There are various ways of solving for the electron wave-functions in such a system. In the method of *augmented plane waves* (APW's) one tries to match travelling plane waves in the interstitial region to sets of spherical harmonics and radial functions satisfying the Schrödinger equation inside each atomic sphere. The matching conditions are now defined over the spherical boundary surfaces, and are therefore much simpler than for a polyhedral unit cell. In *Korringa's method*, or the *method of Kohn and Rostoker* each ion sphere is treated as a source of scattering for the plane waves in the interstitial space, and one looks then for a combination of such plane waves that will stay steady, self-consistently, for this scattering. The 'best' such combinations can be chosen by the application of a variational formula.

In the hands of experts, both these methods can be made to yield good results, consistent with one another and with experiment. The mathematical formalisms are complicated, but the separation of the volume of the crystal into two regions, with appropriate representations of the wave function in each region, is rather nice in principle. This is something like the physical reality.

But there is another extremely elegant method, which is both convenient in use and illuminating in principle. It is a little more subtle than those methods which we have discussed, and therefore needs more careful explanation. We recall a fundamental theorem of wave mechanics, which states that different solutions of the Schrödinger equation must be orthogonal† to one another.

Our conduction electron must be in a state which is orthogonal to all the states of the electrons in the closed shells of the ions, otherwise the conditions

---

† This is the mathematical counterpart of the word 'different' or 'independent' when applied to electron wave functions. If ψ_1 and ψ_2 are truly different solutions of the Schrödinger equation, each capable of being separately occupied by an electron, then the integral of the product $\psi_1 \psi_2$ over space must vanish. In practice it means that the different wave functions must have different numbers, or different arrangements, of nodes.

of the exclusion principle would not be properly satisfied. Herring suggested (in 1940) that one ought to make sure that the wave function had this property from the very beginning before trying to make it satisfy the Schrödinger equation. He proposed a neat technical trick for this. One takes a plane wave with a definite **k**-vector, $\exp(i\mathbf{k}.\mathbf{r})$, and one combines it with an appropriate mixture of wave-like states constructed out of the ion 'core' wave functions—the mixture being chosen so as to be exactly orthogonal to just these core states. This new function is called an *orthogonalized plane wave* (OPW). Then we write down the lattice potential (essentially of muffin-tin form), and calculate what combinations of OPW's (i.e. corresponding to different values of **k**) best satisfy the Schrödinger equation.

This all sounds most artificial—but it turns out that only a very few different OPW's are needed to get quite close to the answer. The point is that each OPW is a very good representation of the sort of wave-function that an electron must have in a metal. The 'core' wave functions are rather compact functions, which vanish outside the closed shells of the ion. Thus, in the interstitial region an OPW looks very like a simple plane wave. But in the core it can have plenty of oscillations and enough nodes to look very like the wave function for the next atomic state above the closed shells.

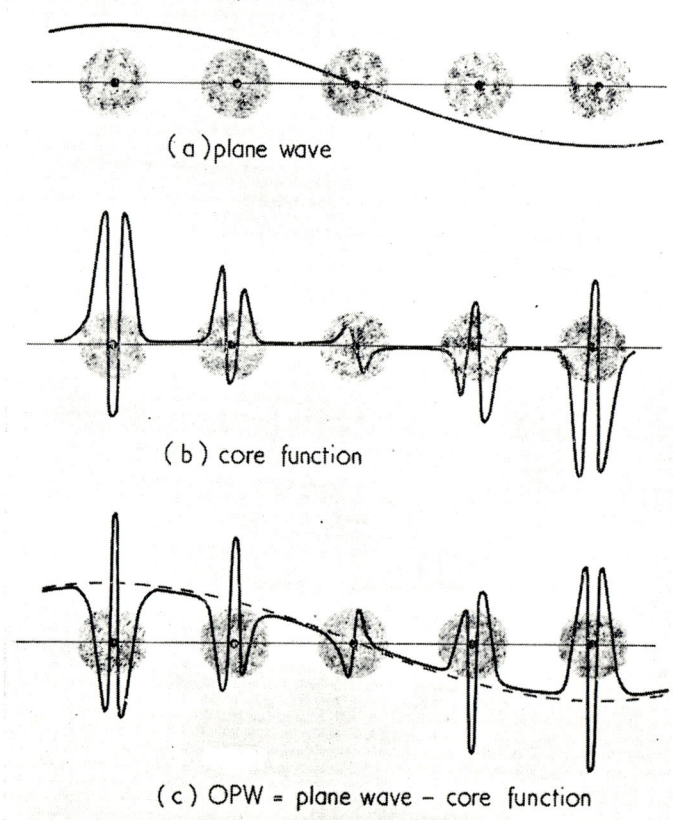

Fig. 30. Synthesis of an orthogonalized plane wave.

7. The effective potential

Out of all these calculations there arises a rather remarkable result. In the end, the conduction electrons look much more free than we had any reason to expect when we started on the problem. The ionic potential is a very strong local potential, with a singularity like Ze^2/r near the nucleus. The Fourier component of this potential at a wavelength equal to the lattice spacing, must be many electron volts, so that we should expect very large energy gaps and strongly distorted Fermi surfaces. All the divalent metals ought, one feels, to have been semiconductors.

Yet in many cases the calculated energy surfaces are not very different from spheres, and the energy gaps are usually only one or two electron volts. In Al, for example, Heine calculated that a single OPW whose energy is nearly the same as that of a free-electron of the same **k**-vector, was a good representation of the solution of the Schrödinger equation, except at points very near a zone boundary. From an *energy* point of view, the free electron model is much better in practice than it appeared at first sight.

We are now beginning to understand the reason for this fortunate circumstance. Calculations of band structure are not simply to find *any* solution for the wave function of an electron in a periodic lattice. There are plenty of these solutions, corresponding to bound states in the ion cores. But these lower states are all filled. We are looking for the 'next' band of states above them. These 'conduction' states must be orthogonal to all the core states, and this puts a serious constraint on the form of the wave-function. It means that the wave function of the conduction electron *must* oscillate rapidly inside the core, and therefore have a rather high kinetic energy. The kinetic energy is enough to balance out most of the potential energy of the core region, so that there is not really a strong barrier to the electron entering or leaving each atomic sphere. This means that each ion does not scatter the conduction electrons very strongly, so that the scattering power of the atomic planes is rather weak.

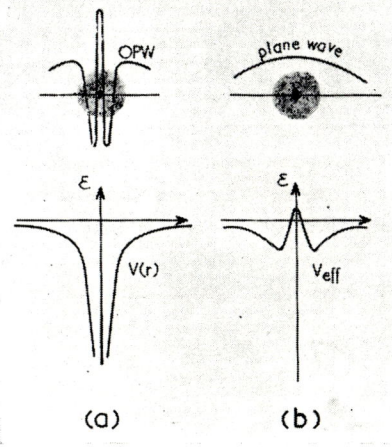

Fig. 31. An OPW in the true potential (*a*) can be replaced by a simple plane wave in the effective potential (*b*).

This argument can be made more or less quantitative. In the OPW method we constructed the final solution of combinations of OPW's, each of which was a plane wave with orthogonalized wiggles in the ion cores. The rules for building up each OPW involve various integrals of wave functions in the core, etc. But these rules can be transformed until, in the end, they can be made to have the same effects as an extra energy term in an ordinary Schrödinger equation. Our problem becomes equivalent to constructing a solution for this modified Schrödinger equation out of simple plane waves.

We have made the orthogonality condition look like a potential. The electron resents the high kinetic energy that it must acquire to enter the core, so that this *pseudo-potential* is obviously repulsive. But the ordinary ion potential is attractive so that these two fields tend somewhat to balance out. The net *effective potential* is thus quite small—just a few electron volts. For such a small potential the solution of the modified Schrödinger equation is very simple; the nearly free-electron method of II § 3 is adequate as a first approximation.

This argument is not rigorous and does not, perhaps, offer the best technical procedure for a precise calculation of band structure. But it shows quite clearly why the free-electron model works so well as a first approximation. We must not insist that the wave function of a conduction electron is just a simple plane wave.—If we do that, the deep potential well at the nucleus of each ion will have an enormous effect on the energy. We must allow the wave function to oscillate rapidly whenever the electron falls into such a well so that it can nearly compensate the local negative potential with a large positive kinetic-energy term. This compensation can be so good that the *energy* of the electron in the lattice is very much the same as if it were perfectly free. The remaining coherent scattering by the lattice planes can be dealt with as a relatively small perturbation.

8. Screening

One general formal problem remains. What atomic potential should we use in our band-structure calculation? How can we allow for the fact that our array of ions is bathed in a dense gas of electrons, which strongly interact with each other and with the lattice? This is a *many-body problem*, for the electrons are identical particles and must be treated collectively—all 10^{23} of them. It has been a challenge to theoretical physics for a generation, and has not been solved completely. But we now think we understand the major phenomena for a free-electron gas (in jellium, of course) even if we have not yet succeeded in putting such a gas into a real lattice of metallic ions.

Happily, our intuitive notions are being confirmed. The idea of a Fermi surface, with the classification of states by their **k**-vectors is still valid. The low-energy excitations of the whole system can be described by a scheme which is equivalent to statements like 'the change of energy when we take an electron out of the state **k**, below the Fermi surface, and put it into **k**′, above the Fermi surface, is given by

$$\delta \mathscr{E} = \mathscr{E}(\mathbf{k}') - \mathscr{E}(\mathbf{k}) .' \qquad (12)$$

Or we can say 'the change of momentum of the whole gas in this transition is exactly $\hbar \mathbf{k}' - \hbar \mathbf{k}$.' Or 'the velocity of the excitation is $\mathbf{v}(\mathbf{k}) = (1/\hbar)\, \mathrm{grad}_{\mathbf{k}}\, \mathscr{E}(\mathbf{k})$'. We can go on thinking of the electrons as nearly independent particles each with charge *e*. The only effect of the strong coulomb interaction between them is to modify the form of $\mathscr{E}(\mathbf{k})$, which is now not quite equal to $\hbar^2 k^2/2m$.

The reason for this is that the electron gas itself acts as a very efficient electrical screen. Put a point charge into the gas—say a negative charge, $-Z|e|$. This will repel electrons. As they move with high velocity (remember the Fermi energy!) past the charge, they will be deflected from it. But there is a background jelly of positive charge. In the region round our foreign body the

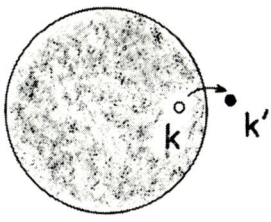

Fig. 32. Excitation of a Fermi gas.

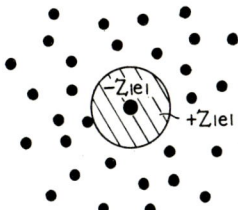

Fig. 33. A negative charge screened by an electron gas.

electron gas is now not sufficiently dense to neutralize the positive jelly. In fact, since our system must be neutral overall, Z electrons will be driven away, and in the neighbourhood of $-Z|e|$ the total excess positive charge on the jelly will be exactly $+Z|e|$. Looked at from afar, the field of our foreign charge is almost exactly compensated by this positively charged cavity in the electron gas.

To calculate this effect in detail, in a self-consistent way, is a formidable problem. But one can show that the field of the foreign charge will be almost negligible at a distance of a few lattice spacings. Instead of $-Z|e|/r$, we shall observe something like

$$\frac{-Z|e|}{r}\exp(-r/l) \tag{13}$$

where l, the *screening length*, is of the order of the interatomic distance.

This argument holds for any charge immersed in a gas of free electrons. We can apply it to the ions of the crystal lattice, and suppose that the strong Coulomb field of each ion is screened by the gas of conduction electrons. The potential seen by any one electron moving through the lattice is thus not nearly as great as we should calculate on the basis of 'bare' ions alone. Here is another reason why the actual effects of the lattice on the ions—the energy gaps etc.—are much smaller than we thought at first.

One can even apply the same analysis to each conduction electron, as if it, too, were a foreign body largely screened by the electron gas in which it is immersed. Each electron is surrounded by a shadowy sphere, into which other electrons do not easily penetrate. The positive charge of the jelly in this sphere equals the charge on the electron, so that the whole object looks electrically neutral. It moves through the volume with nearly the same kinetic energy as a bare electron. There is just the small correction to the free-electron formula because of the difficulty of keeping ' other ' electrons out of the shadow region when our chosen electron is moving rapidly. The interaction between two such excitations is now quite small. It will be a residual electrostatic field like (13), which can only act at distances of the order of the screening length l. The powerful, long-range Coulomb interaction has been cut down to a very local interaction that can easily be treated by perturbation methods and is not very important.

Nevertheless, after all this, the choice of atomic potential is still very uncertain. We have not discussed at all the *exchange* effects, associated with the special interactions between electrons of the same spin. All these band-structure methods, for all the mathematical ingenuity and hours of computer time that have been spent on them, have not really succeeded in giving us good *a priori* knowledge of the electronic structure of metals. When I have said that a method is good, I have meant that its results have turned out to be in good agreement with an already existing experimental determination of the shape of a Fermi surface. The methods for such determinations will be discussed in IV and V.

Part IV. The Properties of Real Metals

1. The story so far

The theory of electrons in metals starts with a gas of free electrons, where the Fermi surface is a perfect sphere. But soon we introduce the crystal lattice. The sphere is carved along the zone boundaries, energy gaps appear, the Fermi surface is divided into several sheets and becomes horribly distorted. But we learn the significance of 'holes' in energy bands, and unmask the mystery of positive Hall coefficients. Hoping then that we might derive $\mathscr{E}(\mathbf{k})$ from first principles, we embarked upon arduous calculations of the band structure. These attempts failed; we never seemed able to calculate the energy gaps with sufficient accuracy. Everything depended on the atomic potential, which was clouded in the fog of correlation and exchange effects (the many-body problem). Our only thread of hope was a tenuous argument suggesting that the effective potential was really rather small, and that the Coulomb interaction between the electrons could be treated as a screening effect. But so far we have made little progress in our quest, for we have not yet predicted, or calculated, or worked out, any exact number for any measurable property of any particular metal.

Now read on...

2. Complex crystal structures and zones

The simple cubic lattice is convenient as a model for the illustration of the general principles of the electronic structure of metals, but no real metal crystallizes in this arrangement. As we saw in I, § 7, there is a preference for more closely-packed structures, with eight or more nearest neighbours for each ion. The very first practical problem in the study of a particular metal is, therefore, to determine its crystal structure and work out the geometry of the corresponding Brillouin zone.

The crystal structure is obtained, of course, by x-ray diffraction. To build up a Brillouin zone we have only to think about the different sets of lattice planes that can give Bragg reflections. Each such set will have a characteristic direction, represented by a unit vector $\mathbf{n}$ in the direction of the normal, and a characteristic spacing d. In $\mathbf{k}$-space we start out from the origin in the direction $\mathbf{n}$, and measure out a length π/d. At this point we erect a plane normal to $\mathbf{n}$. This will be a possible zone boundary. In the standard nomenclature of the subject, the zone boundary is the perpendicular bisector of the *reciprocal lattice vector* **g** = $(2\pi/d)$ **n**:

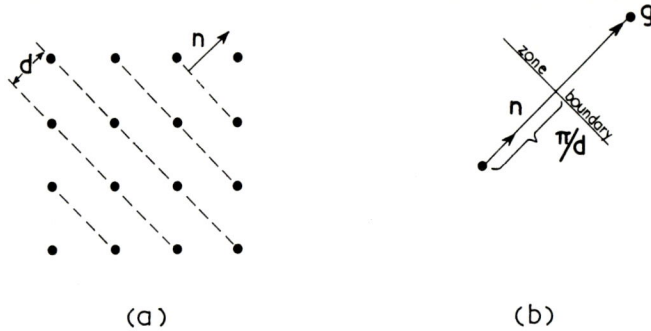

Fig. 34. (a) Lattice planes, spaced d apart give rise to (b), a zone boundary.

For example take the commonest type of metallic structure—the face-centred cubic lattice. This has crystal planes parallel to the x, y, z axes of the cube, spaced $\tfrac{1}{2}a$ apart. There are, therefore, zone faces at $\pi/a\,(2, 0, 0)$ etc. corresponding to these planes. But there are also some well-packed planes normal to the diagonals of the cube (fig. 35 a). For these **n** is in the (1, 1, 1) direction

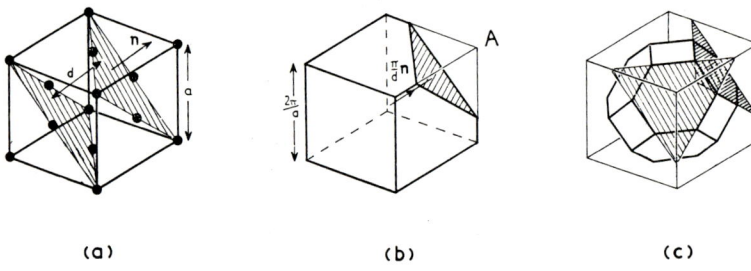

Fig. 35. (a) $(1, 1, 1)$ planes of the face-centred cubic lattice, (b) the corresponding zone boundary, (c) the complete Brillouin zone.

and their spacing is $a/\sqrt{3}$. Thus, the corresponding zone boundary will be at a distance $\sqrt{(3\pi)}/a$ from the origin in **k**-space, and parallel to the diagonal plane of the cube (fig. 35 b). Such a zone boundary cuts the corner off the cube generated by the first set of zone boundaries. There will be eight such sets of diagonal planes corresponding to the eight corners of the cube, and these will also intersect one another, as shown in fig. 35 (c). Thus, the Brillouin zone is rather complicated—it is a tetrakaidekahedron, with eight hexagonal faces and six square faces. Those who study the structure of metals become very familiar with this figure; Pippard (1957) has shown that it makes a most elegant lamp-shade.

This construction exemplifies another basic property of Brillouin zones. The reciprocal lattice vector **g** = $(\pi/a)\,(1, 1, 1)$ actually takes us to A, the corner of the cube in **k**-space in fig. 35 (b). The zone boundary bisects this vector. If we had centred our zone at A, then the same plane in **k**-space would have been one of its faces—the bottom, left, back face instead of the top, right, front one. In fact, the hexagonal faces of our tetrakaidekahedron centred on O and A fit exactly together. Other replicas of the zone can be joined on to the square faces, and the whole fitted together without voids or steric hindrance. The zone of the face-centred lattice can be thought of as the unit cell of a lattice in **k**-space—in fact, the *reciprocal lattice* of our original structure. If one looks at this in detail, one finds that this new lattice is body-centred cubic. Some metals crystallize in the body-centred structure, so *their* zone is the unit cell of a *face-*centred lattice in **k**-space.

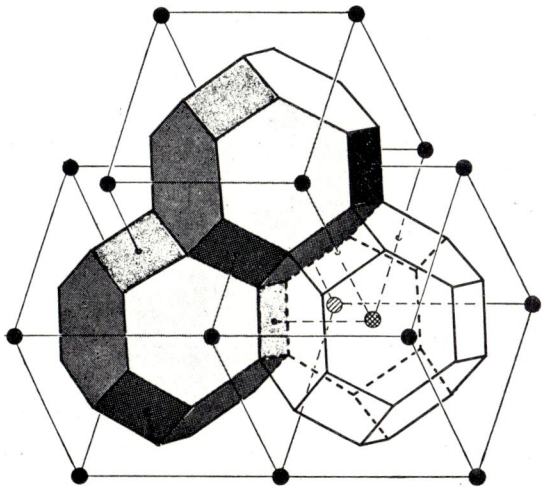

Fig. 36. Stacking FCC zones as unit cells of a BCC lattice.

Since there are only a few different types of structure into which metals normally crystallize, the labour of constructing zones for all of them has been done once for all, and we need only refer to the standard books for details. Nevertheless, this is not quite a trivial problem in such cases as Bi, where the structure is geometrically much more complicated. We can also go further, in a general way, to discuss the symmetry allowed to the wave functions at various points inside the zone without having to introduce any specific form of atomic potential. Consider, for example, the region at the centre of a hexagonal face in the zone of the FCC lattice. It is obvious that the zone has three-fold symmetry about the axis though this point (fig. 37); it is no more than

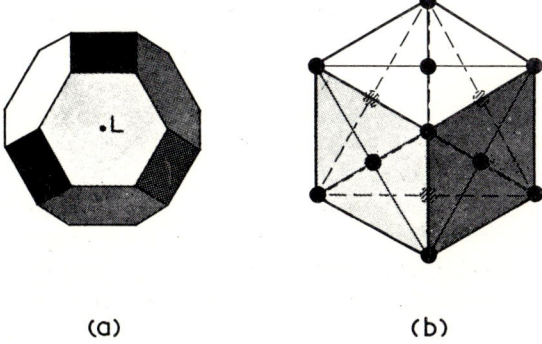

(a) (b)

Fig. 37. (a) Three-fold symmetry of Brillouin zone about L corresponds to, (b) symmetry of FCC lattice about a cube diagonal.

the symmetry of our original face-centred cubic lattice for rotation about a cube diagonal (fig. 37 b). Thus, when we come to construct energy surfaces our function $\mathscr{E}(\mathbf{k})$ must have this same symmetry†. We can learn a lot about $\psi_\mathbf{k}$ and $\mathscr{E}(\mathbf{k})$ at such points just by studying the symmetry properties of the lattice.

The systematic algebraic technique for the complete exploitation of such symmetry properties is called *group theory*, and is an essential tool for the theoretical physicist in this field. It can guide us to the form of the solution before we even consider the sordid details of atomic potentials, and enables us to squeeze the last drop out of an actual calculation. If one wishes to calculate $\mathscr{E}(\mathbf{k})$ at an arbitrary point in $\mathbf{k}$-space by one of the methods discussed in III, one must undertake an extremely lengthy numerical computation. If one is content with values of $\mathscr{E}(\mathbf{k})$ along symmetry lines in the zone (for example), the line from the origin to the centre of the hexagonal face in fig. 37), one can use group theory to shorten and simplify the calculation, and make it worth while.

3. THE ALKALI METALS

The monovalent metals are, in principle, the very simplest ones to study, and have attracted the greatest theoretical attention. What could be easier than a calculation of the electronic structure of Li, with its one valence-electron outside a helium-type closed shell? The fact that the band-structure calculation by different methods have all given different results, even for this elementary case, is only a challenge to our computational skill. Nevertheless, it would be very helpful if we had a direct experimental study of the Fermi surfaces of the alkali metals, so that we might choose the correct set out of the mass of contradictory results.

Unfortunately, most of the experimental techniques require very pure, perfect, single crystals at very low temperatures. They all depend on defining a special direction relative to the crystal axes, by cutting a surface normal to this direction, or applying a magnetic field along it. One can then single out a small group of electrons on the Fermi surface —for example, those travelling parallel to the surface of the specimen—and study their transport properties in detail. A series of such experiments at different crystal orientations can then give enough information to map out the Fermi surface as a whole. But it is obvious that the specimen must itself be a single crystal and in all the techniques it is also essential that the electrons must not be strongly scattered by thermal vibrations or impurities in the metal. The criteria for this will be discussed later, as we consider each technique in detail.

For the alkali metals there are severe metallurgical and chemical problems in the preparation and maintenance of appropriate specimens. These problems are not in principle insoluble but they have so far prevented any detailed study of the electronic structure by direct methods. We must use more roundabout arguments when we try to derive the ordinary transport properties (e.g. electrical conductivity) of the metal from some model of the Fermi surface, and then we adjust the model to make the results agree with experiment.

† In fact, in this case, at the zone face itself the symmetry is even higher—6 fold—because the 'next' zone fits on to the hexagonal face with squares edging hexagons, etc. Near the centre of this face the contours of $\mathscr{E}(\mathbf{k})$ must be very nearly circles.

This method is possible for an alkali metal because we can assume that the Fermi surface is not very far from spherical. The crystal structure is body-centred cubic, whose zone is a dodecahedron. The sphere containing one electron per atom—that is, half the volume of the zone—lies well within the zone. It would require quite large energy gaps on the nearest faces to pull it into contact. One can guess that the Fermi surface is more or less spherical, with some local bulging towards the centres of the twelve facets of the zone. The size of the bulges can be treated as a parameter whose magnitude is to be adjusted to fit the observed properties of the metal.

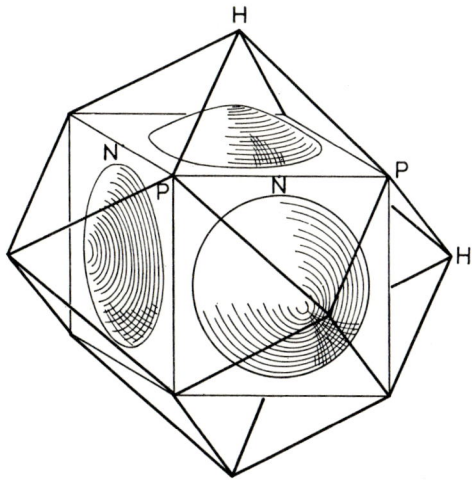

Fig. 38. Zone for BCC lattice, enclosing sphere and cube of half the volume.

A property that has often been used as a test of a band-structure calculation is the *electronic specific heat*. This was discussed in I § 10, where it was shown to be linear in T, with a small coefficient γ. The argument there was for free electrons, and hinged upon the fact that all the electron states are either completely full, or completely empty, except those in the range of energy of width kT about the Fermi level. In a free electron gas the number of states in this range is about $(kT/\mathscr{E}_F)\,n$, where n is the number of electrons per unit volume. More generally, we can express this number as $kT\mathscr{N}(\mathscr{E}_F)$ where $\mathscr{N}(\mathscr{E}_F)$ is the *density of states* function. Thus $\mathscr{N}(\mathscr{E})\,d\mathscr{E}$ is the total number of electron states in the range of energy from $\mathscr{E}$ to $\mathscr{E}+d\mathscr{E}$. One can show quite rigorously that the electronic specific heat must be

$$C_{el} = \frac{\pi^2}{3}k^2 T\,\mathscr{N}(\mathscr{E}_F), \qquad (1)$$

so that it measures quite directly the value of the density of states function at the Fermi level. Since $\mathscr{N}(\mathscr{E})$ can be computed from the form of $\mathscr{E}(\mathbf{k})$, and since C_{el} can be measured to within 1 per cent in the alkali metals, the comparison of theory with experiment is a natural consequence.

As it happens, this is not a very good way of testing a calculation of band-structure. From the uniform density of states in **k**-space, and from the rule (III § 2) $\mathbf{v} = \nabla_\mathbf{k} \mathscr{E}(\mathbf{k})$, one can easily show that

$$\mathscr{N}(\mathscr{E}_F) \propto \int \frac{dS_F}{v} \tag{2}$$

where v is the magnitude of the electron velocity on the element dS_F of Fermi surface, and the integration is over the whole of this surface. A large value of $\mathscr{N}(\mathscr{E}_F)$ may arise for two quite different reasons. It may be that the Fermi surface is very distorted. Where it bulges and approaches the zone boundaries, v tends to become quite small, giving a large contribution to the integral. On the other hand, there may be strong many-body effects, which would reduce the velocity of the electrons at the Fermi level, even if there were no crystal lattice. With only a single measured quantity to be interpreted, one cannot disentangle these two effects. The theoretical comparisons that have been made are almost all meaningless.

4. Transport properties

We might be able to draw some conclusion from the ordinary transport properties of these metals. To do this we need a proper theory of, say, the electrical conductivity of a metal, including the effects of band-structure and Fermi statistics. Let us use the theory of III § 2 for the dynamics of electrons in electric and magnetic fields.

First apply an electric field **E**. This causes each representative point in **k**-space to move at a steady rate:

$$\dot{\mathbf{k}} = (e/\hbar)\mathbf{E}. \tag{3}$$

Suppose that the relaxation time of the electrons is τ; this is the average time for which an electron can be affected by the electric field before being scattered. This means that, on the average, each state is displaced a distance $\dot{\mathbf{k}}\tau$ from its equilibrium (zero electric field) position. In fact, the whole Fermi surface is moved a distance

$$\delta\mathbf{k} = \tau\dot{\mathbf{k}} = (e\tau/\hbar)\mathbf{E} \tag{4}$$

in the direction of the electric field (fig. 39 a).

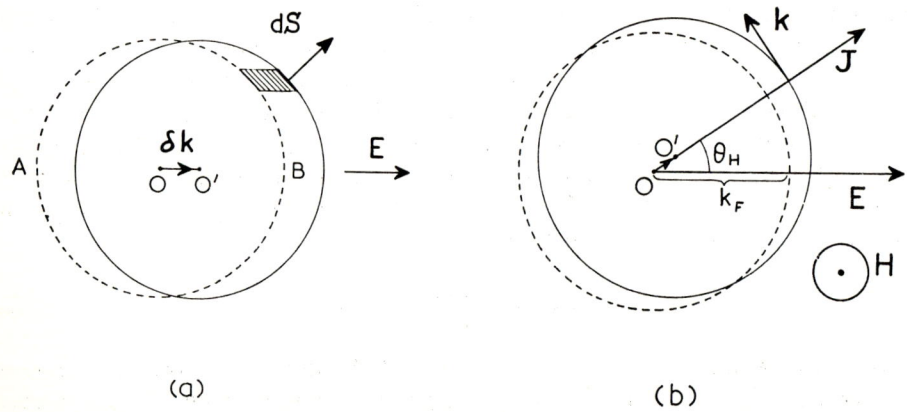

Fig. 39. Displacement of Fermi surface, (a) by an electric field, (b) by an electric field and a magnetic field normal to the page.

A displacement of the Fermi surface is equivalent to an electric current. New states have been occupied in regions of **k**-space where **v** is more or less in the direction of **E**; other states have been emptied in regions where **v** is opposite to **E**. We can calculate this as follows. Consider an element d**S** of Fermi surface. The region that has been occupied is a cylinder with d**S** as base and δ**k** as generator; it will have volume d**S** . δ**k** (this will be negative, of course, in regions which *lose* electrons). Suppose we have a unit cube of metal (i.e. one centimetre, or one metre, cube); one can show† that the density of states in **k**-space is just $1/4\pi^3$. Thus there will be $(1/4\pi^3)$ d**S** . δ**k** electrons added to the Fermi surface here, each with velocity **v**. When we add these all up we get a total electric current density

$$\mathbf{J} = \frac{1}{4\pi^3} \int e \mathbf{v} \, d\mathbf{S} \cdot \delta\mathbf{k}$$

$$= \frac{1}{4\pi^3} \int e \mathbf{v} \, d\mathbf{S} \cdot \left(\frac{e\tau}{\hbar}\right) \mathbf{E}, \tag{5}$$

which is equivalent to a conductivity tensor,

$$\sigma = \frac{e^2 \tau}{4\pi^3 \hbar} \int \mathbf{v} \, d\mathbf{S}. \tag{6}$$

This is the generalization of the formula (2) of I § 4;

$$\sigma = ne^2\tau/m. \tag{7}$$

The two formulae are, indeed, identical for a free-electron gas, as one can verify by putting into (6) the velocity and area of the Fermi surface, and so on. But this is almost a fluke. In the elementary argument we treat each electron in the gas separately, and add them all together, up to a density n. In the quantum-mechanical argument it is really only the electrons at the Fermi surface that count, since they are the only ones that can really be scattered. Although in our derivation we have said that the whole Fermi distribution moves, this is no more than is necessary to keep it uniformly full inside; we could as well say that electrons near A in fig. 39 (*a*) have been carried round to B by the electric field, leaving the deeper layers unaffected.

Unfortunately, it is not easy to calculate τ directly, so that this is still some way from a quantitative argument. Nevertheless, we can see from here explicitly how the resistivity of a metal might depend on the shape of its Fermi surface. If, for example, by reason of zone boundaries, the area of Fermi surface is small, and the electrons there have low velocities, the conductivity is correspondingly small. This is the case in Bi, for example, where all the bands are full, except for a small amount of overlap leaving a few electrons and few holes.

A phenomenon which is more applicable to the case of the alkali metals is the magnetoresistance, since this should vanish identically for a spherical Fermi surface. The proof of this comes quite easily from fig. 39 (*b*). According to III § 2, a magnetic field displaces the electrons at the rate

$$\dot{\mathbf{k}} = \frac{e}{ch} \mathbf{v} \wedge \mathbf{H}. \tag{8}$$

This again may be supposed to act, on the average for the time τ. If an electric field has already been applied so as to produce a displacement of the Fermi

† It follows from the argument of I § 7, and from the two spin states for each electron.

surface, then this magnetic field will tend to rotate the whole Fermi surface, around **H** as axis, with angular velocity $\dot{\mathbf{k}}/k_F$. On the average, the bump carrying the current will be rotated through an angle

$$\theta_H = \tau \dot{\mathbf{k}}/k_F = \frac{e\tau}{ch\,k_F} vH. \tag{9}$$

Thus, there will be this angle between the electric current and the electric field: this is the Hall angle.

From this argument we can derive a formula for the Hall coefficient, which turns out to be exactly the same as we calculated by simple kinetic arguments. The magnetoresistance for a sphere should also vanish, because we can exactly balance the effects of the magnetic field tending to move the displacement away from the current direction by putting **E** at the Hall angle to **J**. But then the displacement itself is due to the component of **E** along **J**, and the relation between these two will be precisely the ordinary conductivity σ. To put it more simply we can exactly balance the net magnetic deflection of the electrons on the Fermi surface by allowing a traverse electric field to build up.

But this argument only works if all electrons on the Fermi surface have the same velocity. If the Fermi surface is distorted, electrons near the zone boundaries have smaller velocities than those on the more distant parts of the surface. One then finds that no single Hall field will exactly cancel the differing amounts of deflection produced by the magnetic field on electrons of differing velocities. The Hall field is a compromise, and various groups of electrons may be lost from the electric current. This means, even in a polycrystalline specimen, an increase of resistance—in low fields proportional to H^2.

Magnetoresistance is interesting because it can give us a direct measure of the deviation of the Fermi surface from a sphere. The magnetoresistance of Na is very small—hence it is believed to be a nearly ideal electron system. In Li it is quite large—hence this metal probably has a rather distorted Fermi surface. The quantitative application of this theory is not very precise, because we have to use arbitrary models for the shape of the Fermi surface, and there are too many parameters that might be adjusted, but the general picture is clear.

5. Phonon drag

Another property that depends on the electronic structure, and is easily measured in a polycrystalline specimen, is the thermoelectric power. In I § 6, §10 we showed how to estimate the order of magnitude of this effect, and in III § 3 it was pointed out that the sign of the thermoelectric power would depend on whether we had electrons or 'holes' as the major carriers. In an alkali metal there is no question but that all the carriers are electrons, so we should expect a negative value of Q_L. This is observed in Na, K, Rb, and Cs, but not, as it happens, in Li. Here is further evidence that the Fermi surface of Li is rather distorted.

Unfortunately the theory of the thermoelectric power is rather complicated. By considering contributions to the currents of heat and electricity from electrons in the range of energy kT around the Fermi level, one can derive a formula

$$Q_e = -\frac{\pi^2 k^2 T}{3\,|e|} \left[\frac{\partial \ln \sigma(\mathscr{E})}{\partial \mathscr{E}}\right]_{\mathscr{E}=\mathscr{E}_F}, \tag{10}$$

where the derivative is taken of " what the conductivity would be if the Fermi level stood at $\mathscr{E}$ ". This quantity obviously depends on the relaxation time of the electrons, which may be a function of their energy. The connection between Q_e and the shape of the Fermi surface is more roundabout, and this quantity has not yet really been very useful as a quantitative index of the electronic structure.

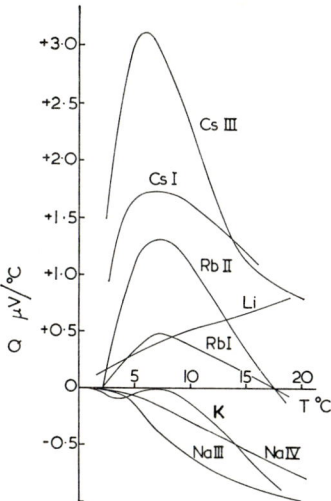

Fig. 40. Thermoelectric power of the alkali metals at low temperatures; different specimens (Na III, Na IV, etc.) give different results, depending on parity (MacDonald).

But the alkali metals exhibit very peculiar behaviour at low temperatures. The thermoelectric power, instead of going linearly to zero as suggested by (10), may become quite large in magnitude. In Na and K it stays negative, but in the other alkali metals there are large positive humps in the curve, as shown in fig. 40. Each curve is characteristic of the metal. This effect is explained as follows:

The conduction electrons in a metal are scattered by the thermal vibrations of the crystal lattice. This is the origin of the temperature-dependent resistance of metals—the part that is proportional to T at ordinary temperatures. The amplitude of the thermal vibrations, hence the amount of scattering of electrons, depends on the temperature, just as in the theory of the lattice specific heat.

But the thermal vibrations of a crystal lattice are not simply independent oscillations of the individual atoms, as if each were isolated from its neighbours. To treat them properly, we have to analyse them into travelling waves, like ordinary sound waves though of much shorter wavelength. These waves are quantized in momentum and energy, and then have properties analogous to the photons of the electromagnetic field. We call them *phonons*. The scattering of an electron by a lattice vibration is then described as " the emission (or absorption) of a phonon by the electron ", just as in the theory of elementary particles.

Now if we have a current of electrons flowing through the metal in a particular direction, there will be a lot of phonons emitted in that direction as the electrons meet the resistance due to the thermal vibrations of the crystal. These phonons will, in their turn, be scattered by impurities, etc., but meanwhile they constitute an extra current of excitations flowing with the electrons. Phonons have a large specific heat—much larger, usually, than the electronic specific heat—so that this current of phonons 'dragged along' by the electrons will constitute a substantial current of heat. Just as in the argument of I § 6, this heat current gives rise to a Peltier effect, and hence is observable in the thermoelectric power of the metal. The effect is only to be seen at low temperatures because otherwise the phonons are themselves too strongly scattered. At still lower temperatures the lattice specific heat becomes small, so the effect again disappears.

But what of the sign of the *phonon drag* contribution? At first sight, one would think that it must be negative, like the charge of the electrons which 'drag' the phonons along. This is not so, for a very subtle reason. In the process by which an electron emits a phonon we say that "Energy is conserved", i.e.

$$\mathscr{E}(\mathbf{k}) = \mathscr{E}(\mathbf{k}') + \hbar\nu: \tag{11}$$

the energy of the electron as it goes from the initial state $\mathbf{k}$ to the final state $\mathbf{k}'$ is diminished by the energy, $\hbar\nu$, of the phonon. We should also like to have a rule that "momentum is conserved". The 'momentum' of a Bloch electron is $\hbar\mathbf{k}$. Correspondingly, the 'momentum' of a phonon is $\hbar\mathbf{q}$, where $\mathbf{q}$ is the wave vector of the lattice wave (of frequency ν, of course). Thus, in the process characterized by (11) we expect to write

$$\hbar\mathbf{k} = \hbar\mathbf{k}' + \hbar\mathbf{q}. \tag{12}$$

This, however, is not the whole story. If we try to derive a formula like (12) we keep finding cases where a transition process is still allowed, and yet where

$$\mathbf{k} - \mathbf{k}' - \mathbf{q} = \mathbf{g}, \tag{13}$$

$\mathbf{g}$ being one of the reciprocal lattice vectors in § 2 above. This is called an '*Umklapp*' *process* (or *U-process*). It seems to be against Nature; but we must remember that $\hbar\mathbf{k}$ is not really the dynamical momentum of the electron in the crystal. Indeed, this 'crystal momentum' is not unique. As we saw in II and III, the state $\psi_\mathbf{k}$ is not simply a plane wave, exp $(i\mathbf{k}\cdot\mathbf{r})$. It contains admixtures of other waves, like exp $\{i(\mathbf{k}-\mathbf{g})\cdot\mathbf{r}\}$. Thus, $\hbar(\mathbf{k}-\mathbf{g})$ is as good as $\hbar\mathbf{k}$ as a label for the crystal momentum of this state. If we make this replacement in (12), we arrive at (13). Another way of thinking about it is in terms of Bragg reflections. An Umklapp process is one in which the electron undergoes a Bragg reflection (change of wave-vector by $\mathbf{g}$) at the moment when it is emitting a phonon.

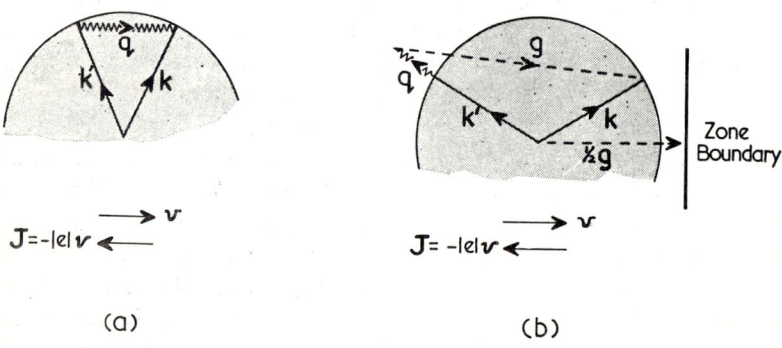

Fig. 41. Electron scattered from $\mathbf{k}$ to $\mathbf{k}'$ emits phonon $\mathbf{q}$, (*a*) N-process, (*b*) U-process.

Now back to the problem of phonon drag. A 'normal' electron scattering process is indicated in fig. 41 (*a*). The electron in **k** is scattered to **k'**, and reverses its component along the current, **v**, of charges. The phonon is emitted parallel to **v**. This gives an ordinary negative thermoelectric power. In a U-process, fig. 41 (*b*), we have to make up a vector **g**. But ½**g** is the distance from the origin to a zone boundary, so that **g** is a large vector compared with the diameter of the Fermi sphere. Thus, **q** tends to be in the opposite sense to **k** − **k'**—the heat current is more or less *antiparallel* to **v**, and we have a positive thermoelectric power.

The balance between these two effects is rather delicate, and the result depends on several factors including the form of the lattice spectrum. But it is particularly sensitive to the shape of the Fermi surface. Any bulging of the Fermi surface towards the zone boundary means that U-processes can occur with smaller values of q, and the thermoelectric power becomes more positive. The curves of fig. 40 can easily be explained if we assume that in Na the Fermi surface is nearly a perfect sphere, in K also it is only slightly distorted, in Rb it is rather more knobbly, and in Cs and Li the Fermi surface comes quite close to touching the zone boundary.

This analysis is confirmed by a rather more complicated calculation of the behaviour of the electrical and thermal conductivities of these metals as functions of temperature. Again, the Umklapp processes play a decisive role, but we need a rather detailed model of the electron wave functions and of the frequency distribution of lattice waves to get good quantitative results.

These attempts to discover the electronic structure of the alkali metals are obviously not very precise, but they do show what can be deduced from indirect evidence. It is nice to know that these qualitative results are consistent with each other, and with at least one direct theoretical calculation of the band structure of these metals, using the Kohn and Rostoker method. Our original choice of Na as the ideal free-electron metal is confirmed.

6. The anomalous skin effect

The 'noble' metals, Cu, Ag and Au, are easily obtained in very pure and perfect single crystals. They are therefore ideal for the application of more advanced techniques for the study of the Fermi surface, etc. Although they are monovalent, their electronic structures turn out to be quite complicated, and are now understood in great detail. The first "experimental investigation of the Fermi surface of a metal" was the classical work of Pippard, who studied the anomalous skin effect in Cu. His conclusion—that the Fermi surface of Cu is very distorted and almost certainly touches the zone boundary—was a great stimulus to further work in this field, as it showed the inadequacy of the free-electron model even for this simple metal.

The method depends upon the following phenomenon. Consider the actual surface of a piece of metal, with a high frequency (r.f.) field in the space outside. It is well known that this field induces currents in the metal which prevent penetration of the field through the specimen. But the depth of penetration—the *skin depth*, δ—depends on the frequency ω of the r.f. field and on the electrical conductivity, σ_0, of the metal. It is easy to show, from classical electromagnetic theory, that one should have

$$\delta_{cl.} = 1/\sqrt{(2\pi\omega\sigma_0)}. \tag{14}$$

But suppose one does the experiment at very high frequencies on a very pure specimen at very low temperatures, where the conductivity is very high. The skin depth becomes very small—much less than the mean free path, Λ, of an electron. We can check this by calculating Λ from I § 4, or IV § 4;

$$\Lambda = \tau v = m\sigma_0 v_F/ne^2. \tag{15}$$

Indeed, by increasing σ_0, we have also increased Λ, until $\Lambda \sim \delta_{cl.}$.

The classical theory then fails. The only electrons that can receive energy from the electric field are those which are running parallel to the surface of the specimen. Other electrons travel right through the skin layer, and do not take

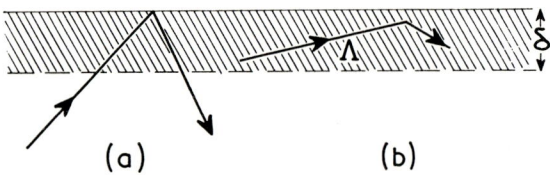

Fig. 42. (a) Electron passing right through the skin layer, (b) electron travelling nearly parallel to surface is scattered within the skin depth.

up enough energy in their rapid transit to affect the conductivity. It is as if only the fraction δ/Λ of the electrons—the ones that stay in the skin depth long enough to make a collision with an impurity—were available. It is as if the conductivity were only

$$\sigma' \sim (\delta/\Lambda)\sigma_0. \tag{16}$$

But the general electrodynamic relation (14) still holds. The skin depth must be established self-consistently so that

$$\delta \sim 1/\sqrt{(2\pi\omega\sigma')} \sim 1/\sqrt{(2\pi\omega\delta\sigma_0/\Lambda)}. \tag{17}$$

We can solve this for the 'anomalous' skin depth:

$$\delta_{\text{anom.}} \sim (2\pi\omega\sigma_0/\Lambda)^{-1/3}. \tag{18}$$

What we actually measure is the surface impedance, but the same result holds; in the anomalous region, this does not depend directly on Λ, but only on the ratio σ_0/Λ. For a free-electron metal, as in (15), (σ_0/Λ) is a constant, dependent on the velocity of the electrons at the Fermi surface but not on the mechanism of scattering. Thus, the anomalous skin effect measures directly a feature of the electronic structure of the metal. Moreover, it is not an average property over the whole Fermi surface. Only electrons with velocity parallel to the boundary of the specimen have contributed. If the Fermi surface is distorted, then we may get different effects for surfaces cut at different orientations to the crystal axes.

A more detailed analysis shows that the quantity which is measured—the surface resistance R_x when the electric field vector is in the x direction, say—is proportional to

$$\{\int |\rho_y| dk_y\}^{1/3}, \tag{19}$$

where the integration is round a belt of the Fermi surface where the electron velocity is parallel to the face of the specimen, and $|\rho_y|$ is the radius of curvature of a section of the Fermi surface in the plane normal to the face of the specimen and parallel to the x axis. It is a purely geometrical parameter of the Fermi surface in **k**-space.

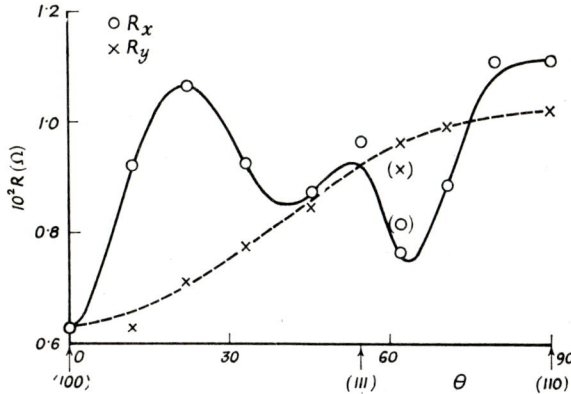

Fig. 43. Surface resistance of Cu in the anomalous region, for various orientations of a single crystal (Pippard).

Figure 43 shows how greatly the surface resistance varies as we change the angle of cut on the face of a crystal of Cu. It is immediately obvious that this metal cannot have a spherical Fermi surface. The problem of 'inverting' a formula like (19) is extremely complicated, but by a process of trial and error one can come to a surface which has something like the shape to fit the observed results. This is shown in fig. 44.

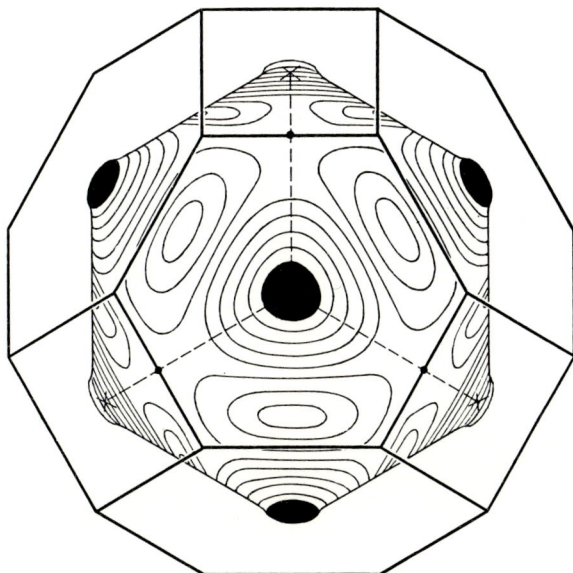

Fig. 44. Fermi surface of Cu, to fit anomalous skin effect (Pippard).

Cu crystallizes in a face-centred structure, whose zone has already been discussed in § 2. Pippard was able to show that the anomalous skin effect could not be explained unless the Fermi surface is so distorted that it comes into contact with the zone boundaries. We find that the sphere has been pulled out into cylindrical projections meeting the hexagonal faces of the zone over a substantial area.

The anomalous skin effect has now been used in the study of several metals. Its advantage is that it gives weight to the major regions of the Fermi surface—the regions that contribute most to the electrical conductivity. The disadvantages are: that it lumps together effects from several bands, making the analysis of polyvalent metals rather difficult; that the measured quantity is not a very simple geometrical feature of the Fermi surface and the labour of adjusting a model to fit the results is considerable; and that it is difficult metallurgically to obtain sufficiently clean, flat surfaces to do a good experiment.

The other transport properties of the noble metals have also been studied, and correlated with Pippard's model. We find, for example, a positive thermo-electric power at high temperatures, and also positive 'phonon-drag' humps at low temperatures. These are consistent with a Fermi surface that touches the zone boundary. However, such indirect evidence is as nothing compared with the direct results obtained by the special techniques using high magnetic fields, to be discussed in Part V.

Part V. Gauging the Fermi Surface

*I'll put a girdle round about the earth
In forty minutes.*—A Midsummer Night's Dream.

1. Magnetic orbits and cyclotron resonance

In Part III, § 2, we discussed the effect of a magnetic field on the state of an electron. We saw that the state vector is changed at the rate

$$\dot{\mathbf{k}} = \frac{e}{c\hbar} \mathbf{v} \wedge \mathbf{H}. \tag{1}$$

The motion of the representative point in **k**-space is thus normal to the direction of the magnetic field. The point will always stay on the same plane in **k**-space. Moreover, the motion is normal to the 'local' electron velocity, **v**, which is proportional to the gradient of the energy function $\mathscr{E}(\mathbf{k})$, and is thus normal to the surfaces of constant energy in **k**-space. This means that $\dot{\mathbf{k}}$ lies in the tangent plane to $\mathscr{E}(\mathbf{k})$ at **k**. The representative point stays always on the same energy surface. We have the following simple rule:

In a magnetic field **H** *the* **k**-*vector of an electron on the Fermi surface traces out the 'orbit' defined by the intersection of the Fermi surface with a plane normal to* **H**.

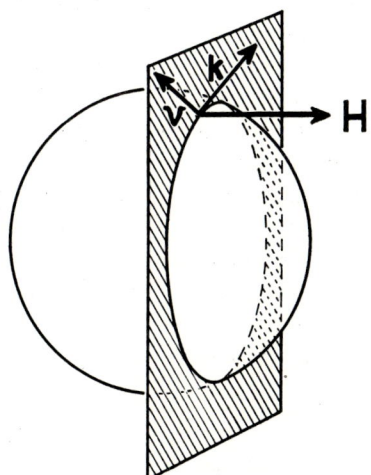

Fig. 45. The magnetic orbit is the intersection of the Fermi surface with a plane normal to the magnetic field.

This elegant theorem is at the heart of all other methods of studying the Fermi surface. A magnetic field does not change the energy of an electron. It merely deflects it into a complicated helicoidal path in space. But in momentum space this path is reflected as an *orbit* which is very easy to define and comprehend. These orbits are directly expressible in terms of the geometry.

of the Fermi surface. They are, we feel, much more than mathematical constructions for the expression of the solutions of differential or integral equations.

But in order to realize such an orbit we must allow the equation of motion (1) to proceed for a sufficient time. For example, the representative point will make a whole circuit of the Fermi surface in the time

$$T = 2\pi/\omega_H = \frac{c\hbar}{eH} \oint \frac{dk}{v_\perp}, \tag{2}$$

where $v_\perp$ is the component of the velocity of the electron normal to **H** and the integral is taken round the circumference of the orbit. Since **k** has now returned to its starting point, the motion is repetitive, with period T, or, as indicated in (2), angular frequency ω_H. This is the *cyclotron frequency* for the orbit. It is easy to calculate that for a free-electron sphere

$$\omega_H = eH/mc, \tag{3}$$

a familiar formula going back to Larmor's calculation of the precession of the motion of classical charges in a magnetic field.

In an ordinary metal at room temperature this effect cannot be observed. The relaxation time, τ, of an electron, due to scattering by impurities and lattice vibrations, is then about 10^{-14} secs. We should need to apply an enormous magnetic field, putting the cyclotron frequency in the optical range, to see a whole cycle between collisions. But we can make $\omega_H \tau > 1$ in an attainable field by using a very pure specimen and going to very low temperatures. The first successful observations of cyclotron resonance in Cu were actually made on native Cu crystals, borrowed from a geological museum! The conductivity of such a specimen at liquid helium temperature may be several thousand times larger than at room temperature. Even then, one needs magnetic fields of tens of thousands of oersteds, and the cyclotron frequency is in the microwave range.

The most obvious experiment is the direct observation of the cyclotron frequency by resonance with a microwave field. This is a well-known effect in semiconductors, where it has been used for some time in the study of band structure, effective masses of carriers, etc. But in a metal there is a difficulty. Because of the skin effect, the microwave field would not penetrate into the metal to a sufficient depth to couple with the electron motion. Each electron is screened by the dense gas of other electrons.

This difficulty can be circumvented. The trick (due to Azbel' and Kaner) is to put the magnetic field *parallel* to the surface of the metal. Then the electrons will move in circular or helical paths, as in fig. 46. Most of each

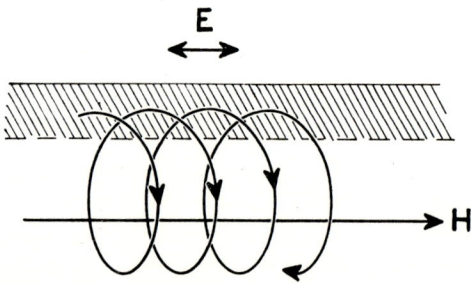

Fig. 46. Cyclotron resonance in a metal, with magnetic field parallel to the surface.

path is below the skin depth, and therefore does not see the oscillating electric field. But if, each time it makes a circuit, an electron arrives at the surface in phase with the electric field, it will rapidly absorb energy from that field. Thus, a resonance can be obtained at the cyclotron frequency of those electrons. Moreover, the resonance should also occur at harmonics of ω_H, where the electric field has made 2, 3, etc. oscillations before the electron returns to the surface. The procedure in practice is to keep the oscillation frequency fixed, and to increase the magnetic field. We see a beautiful series of peaks, equally spaced in the variable $1/H$.

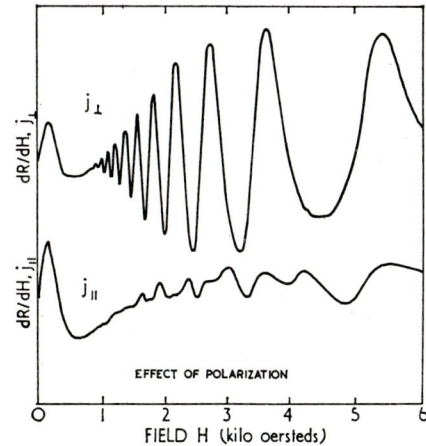

Fig. 47. Cyclotron resonance in copper (Kip 1960).

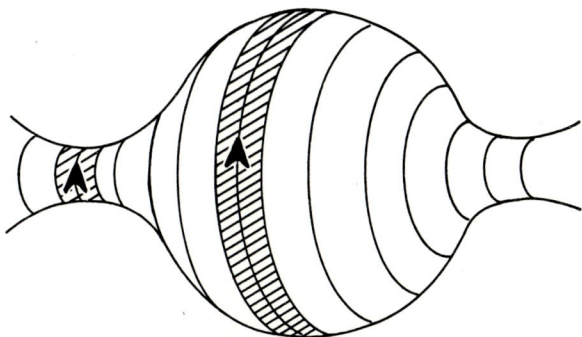

Fig. 48. Extremal orbits.

The interpretation of these oscillations is not quite obvious. For a spherical (or ellipsoidal) surface all electrons have the same cyclotron frequency for a field in a given direction, and there is no problem. But for a general Fermi surface different sections will have different values of ω_H, and the resultant will be a mixture of contributions from all parts. However, if we think of the magnitude of the contributions from different sections, we can see that there will usually be a rather large effect from those sections where the cross-sectional area of the surface is an extremal. The cyclotron frequency around here will be stationary, and so there will be a large belt where it is nearly constant. The electrons from this region will dominate the effect, so that the observed oscillations in the resonance curve can usually be ascribed to *extremal orbits* on the Fermi surface. For example, in the case of Cu, if the field is along (1,0,0) we observe only one resonance, because there is only one extremal cross-section, but in the (1,1,1) direction we can see both a large orbit, drawn round the ' belly ' of the surface, and a smaller orbit round the circumference of the region of contact with the zone boundary.

2. The repeated zone scheme

Consider, now, another orbit in Cu. Put on a field in the (1,1,0) direction so that the central section of the Fermi surface will cut through the regions of contact (fig. 49). Consider an electron starting, say, at A. Under the influence of the magnetic field it will move along the section, until it reaches B, on the zone boundary. What happens then? This is a theoretical question of fundamental importance, whose answer dominates phenomena in magnetic fields.

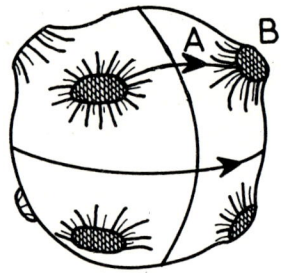

Fig. 49. Orbits on the Fermi surface of copper.

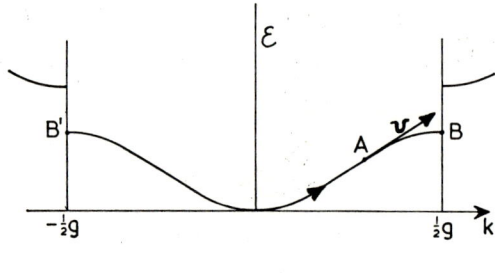

Fig. 50.

Let us go back to our one-dimensional model. Let us apply an electric field, so as to urge the representative point along the curve $\mathscr{E}(\mathbf{k})$ at a steady rate. What happens when we reach the zone boundary B (fig. 50)? The velocity becomes zero, because $\mathscr{E}(\mathbf{k})$ has a maximum there. But the acceleration will still continue. The electron cannot acquire enough energy to jump up into the next band. Into what state does it go?

Well, what state is it in at this point B? According to the analysis of II, § 2, it is in a standing-wave state, such as

$$\psi_+(\tfrac{1}{2}g) = \sqrt{2} \cos \tfrac{1}{2}gx = \frac{1}{\sqrt{2}} \{\exp(\tfrac{1}{2} igx) + \exp(-\tfrac{1}{2} igx)\} \qquad (4)$$

(the state ψ_- belongs to the next band). This state is a combination of two plane waves, of equal amplitude, one from $k = \tfrac{1}{2}g$—i.e. from B itself—the other from $k = -\tfrac{1}{2}g$, the 'other' zone boundary. But suppose we had looked at the electron state in this band at this 'other' zone boundary—at B'. Obviously, by reversing the sign of $\tfrac{1}{2}g$, we get at once:

$$\psi_+(-\tfrac{1}{2}g) = \frac{1}{\sqrt{2}} \{\exp(-\tfrac{1}{2}igx) + \exp(\tfrac{1}{2}igx)\} \qquad (5)$$

—*exactly the same as at B*. The points B and B' are equivalent in our diagram. They refer to the same state. The electron has the same wave function at $k = \tfrac{1}{2}g$ as at $k = -\tfrac{1}{2}g$.

For the counting of states this raises no problem. We agree to include all 'allowed' values of k between B and B', but only include one of the end points; the distribution of allowed points is so dense that the inclusion of both points would make a negligible error. But we can see now exactly what happens as the representative point is urged up to B and through the looking glass. It jumps to B', and comes down the $\mathscr{E}(\mathbf{k})$ curve from there. Its velocity passes through zero, to negative values, so that the electron will seem to be moving *against* the electric field that has been 'accelerating' it. Yet we should interpret this, quite naturally, as a Bragg reflection, brought about as we tried to draw the electron through the critical point in wavelengths at which the Bragg condition is satisfied.

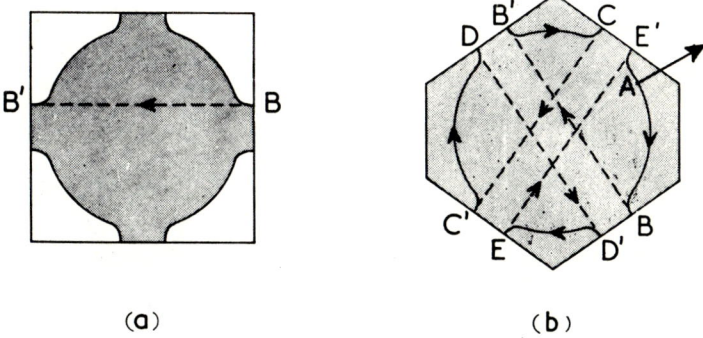

Fig. 51. (*a*) Electron 'jumps' from B to B', which are equivalent. (*b*) A complete orbit, with 'jumps', in a reduced zone.

In three dimensions in a magnetic field just the same will occur. In a square zone, fig. 51 (*a*), the electron state at B is exactly the same as at B', the point on the zone that one reaches by making a translation of exactly one reciprocal lattice vector, $\mathbf{g}$, in $\mathbf{k}$-space. The 'jump' is equivalent to one reflection in the set of lattice planes, normal to $\mathbf{g}$, which defined this zone boundary at B.

62   J. M. Ziman

In the case of the (1,1,0) section of Cu, fig. 51 (*b*), we make a series of jumps: from *B* to *B'*, then along a piece of orbit to *C*; from *C* to *C'*, and along another section or orbit—and so on. The 'jumps' are instantaneous, so the 'orbit' is just the four segments of solid line round the section of the Fermi surface, excluding zone boundaries, in the alphabetical order of the labels.

This is the formal solution to our problem. But it can be made much more elegant geometrically. Instead of making our representative point jump back by a whole reciprocal lattice vector each time it reaches a zone boundary, let us translate the whole zone itself, and the Fermi surface with it, through the vector $-\mathbf{g}$, so that the correct region is waiting just *through* the boundary. In one dimension we get a simple periodic curve, with smooth maxima at the zone boundaries and minima at the 'zone centres' (fig. 52). An electric field would drive k steadily along this curve, and smoothly through the boundary point, which would not be noticed except as a point of zero velocity. In two dimensions (i.e. in the section of a three-dimensional zone) we have to put B' at B, then

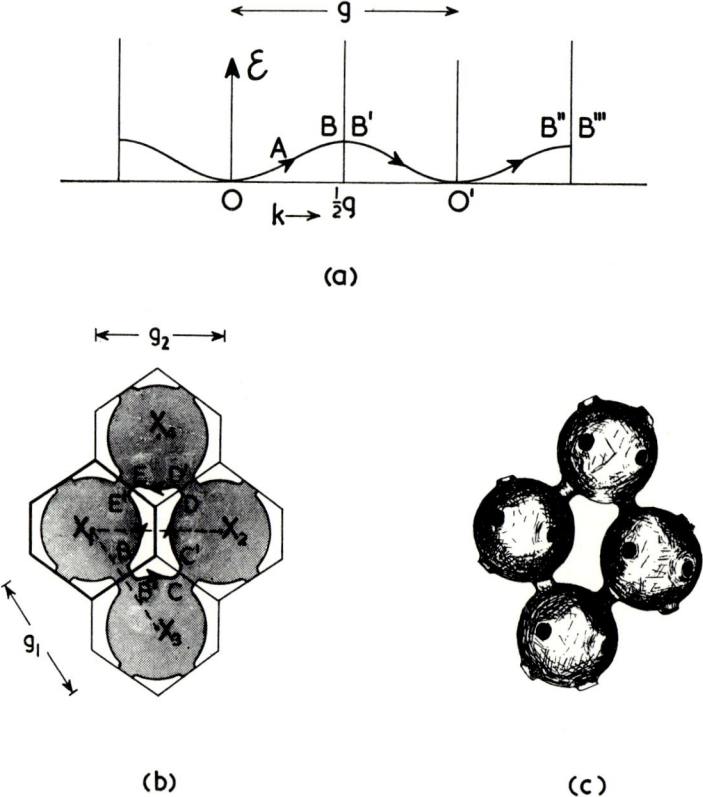

Fig. 52. Orbital continuity restored, (*a*) in one dimension, (*b*) in two dimensions (*c*) in three dimensions, by using repeated zone scheme.

repeat the figure again, with C' at C, and so on. The result, fig. 52 (*b*), is now a figure in which the orbit is a *closed curve*—known fancifully in the trade as a 'dog's bone'. In three dimensions this is equivalent to putting together a series of figures like fig. 44, joining the contact regions across the zone boundaries

This complicated geometrical object, fig. 52 (c), with its more or less spherical 'bellies' joined by eight necks through the cube diagonals, is the Fermi surface of Cu in the *repeated zone scheme*.

This construction is actually more than a convenient geometrical device for making the orbits seem continuous. The Brillouin zone is the unit cell of the reciprocal lattice in **k**-space (IV, § 2) and can, therefore, automatically be repeated exactly as required to fill that space. The **k**-vector of a state is not uniquely defined (IV, § 5 and (4) and (5) above) and can be taken to be arbitrary up to the addition of any vector of the reciprocal lattice. Thus in fig. 52 (b) a state with a representative point at X_1, say, in the 'original' zone could be at the equivalent points, X_2, X_3, etc. in any of the repetitions of this zone obtained by translation through the reciprocal lattice vectors $\mathbf{g}_1$, $\mathbf{g}_2$, etc. One can show that $\mathscr{E}(\mathbf{k})$ is a continuous periodic function in **k**-space, with the period of the reciprocal lattice, so that the movement across the zone boundary is quite smooth and uneventful. 'Belly', 'neck', and 'dog's bone' orbits are all equally permissible. It does not matter that the first is confined to the interior of a zone, the second lies in the plane of a zone boundary, and the third enters four different zones in succession. From a mathematical point of view, our multiply-connected Fermi surface is a continuous object repeated endlessly to infinity, without joints or seams.

Cyclotron resonance allows us to look at the closed orbits corresponding to extremal sections of our repeated Fermi surface. Unfortunately, the values of ω_H for extreme orbits are not very useful as primary information about the shape of the Fermi surface. As we see from (2), ω_H contains the electron velocity, which depends on the spacing of the energy surfaces in **k**-space, not on the geometry of a particular surface. But when, as in Cu, we have found the shape of the Fermi surface itself, very exactly, by other methods, the cyclotron frequencies provide us with derivatives of $\mathscr{E}(\mathbf{k})$ in **k**-space, and thus a means of extrapolating this function for some distance away from the Fermi surface. Again, bandstructure calculations can provide estimates of ω_H that can be checked in detail by experiment.

3. Magnetoresistance

In the discussion of cyclotron resonance we have only considered closed orbits, such as those around the 'belly' of the Fermi surface in Cu, or the 'dog's bone' orbit, which is actually a 'hole' orbit since it encloses an empty region in **k**-space. But through a multiply-connected Fermi surface we can usually make a section whose edge is not a closed curve at all. An obvious case is shown in fig. 53 (a) where we slice the system along one of the (1,1,1) axes; the motion of the representative point never returns to its starting point. There are more complicated sections which have this property. For example, one can prove by simple logic the topological theorem that any section, such as fig. 53 (c), which has 'electron' orbits in one region and 'hole' orbits in another region will have an *open orbit* separating the two regions.

When we apply a magnetic field in a particular direction, and measure the d.c. conductivity, we have to think of the contributions of all the electrons on the Fermi surface. This means that we must look at all the sections that can be made normal to this field axis. It is very likely, then, that, on some of these sections there will be open orbits. In a low magnetic field this distinction is unimportant. For a closed orbit where $\omega_H \tau \ll 1$, the electron never completes more than a fraction of its circuit before it is scattered. The magnetoresistance is then an average, over the whole Fermi surface, of local curvatures, electron velocities, etc., regardless of whether the small areas through which the electrons move before they are scattered are parts of open or closed orbits. We have already noticed (IV, § 4) that some useful information can be obtained from the *low-field magnetoresistance* but it needs careful interpretation.

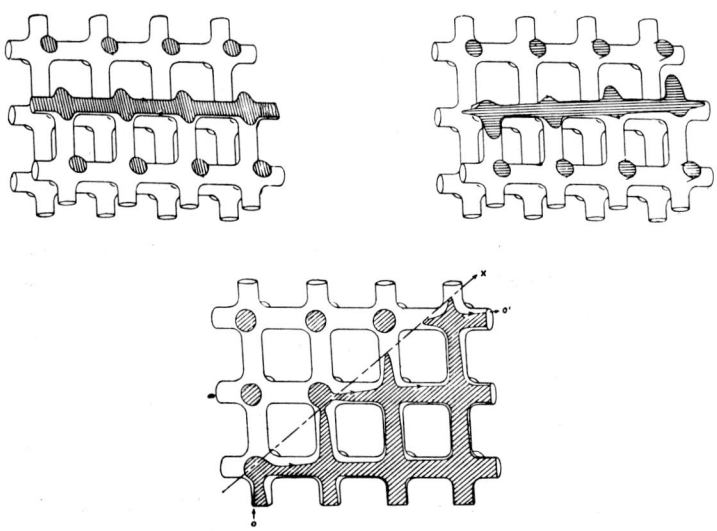

Fig. 53. Demonstration of topological principle that an open orbit always separates regions of 'electron' orbits and 'hole' orbits. These figures arise by cutting a cubic scaffolding of pipes by planes at various angles (Chambers 1960).

But when we go to high magnetic fields we notice an important difference between open and closed orbits. In a closed orbit, if $\omega_H \tau \gg 1$, the electron makes many circuits before it is scattered. From the point of view of calculating electrical conductivity, the velocity of this electron is not the velocity at some special point on the Fermi surface; it is an average velocity taken round the whole orbit. One can see that the component of average velocity normal to **H** will tend to zero as H increases. If **H** is the z-axis, and if we measure the conductivity in the x-direction, then we expect it to tend to zero as H tends to infinity. One can easily show, for a closed orbit, that

$$\sigma_{xx} \to A/H^2, \text{ as } H \to \infty, \tag{6}$$

where A is a constant that depends on the nature of the scattering processes, the geometry of the orbit, etc.

For *open* orbits look at the sections in fig. 53. These are normal to the magnetic field, and there is no symmetry to ensure that the average component of velocity of the electron in this plane should vanish. If the x-axis is the direction along which the orbit stretches to infinity, we certainly expect v_y to remain finite in the strongest magnetic field. The corresponding component of conductivity would then tend to a constant:

$$\sigma_{yy} \to B \text{ as } H \to \infty. \tag{7}$$

Looking at these two formulae, we might suppose that the *resistance* of a metal transverse to a magnetic field would tend to infinity if all the orbits were closed, but would tend to a finite limit if there were some open orbits in the appropriate directions. Unfortunately (and much to one's mental confusion!) the observed behaviour is just the converse.

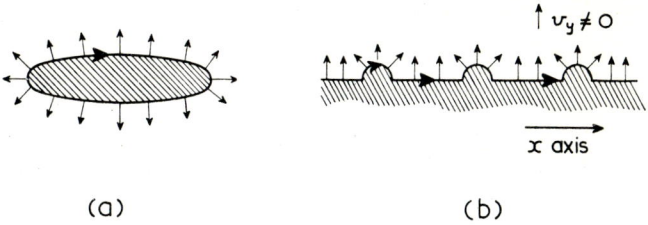

Fig. 54. (*a*) On a closed orbit the average velocity in the magnetic plane is zero. (*b*) On an open orbit, some components of velocity need not vanish.

To get the correct answer, we write down the tensor relations between current and field. In the *xy*-plane we have the following equations.

$$J_x = \sigma_{xx} E_x + (1/RH) E_y \qquad (8)$$
$$J_y = (-1/RH) E_x + \sigma_{yy} E_y.$$

The off-diagonal terms express the Hall effect, through the Hall coefficient R (see I, § 5, IV, § 4). This coefficient may not be quite the same in a very high magnetic field as it is in low fields, but the electric field produced is still at right angles to the electric current, and approximately proportional to H.

To discuss resistivity we must solve these equations for E_x and E_y as functions of the components of current J_x, J_y: e.g.

$$E_x = \frac{1}{\sigma_{xx} + 1/(R^2 H^2 \sigma_{yy})} J_x + \frac{RH}{1 + R^2 H^2 \sigma_{xx} \sigma_{yy}} J_y. \qquad (9)$$

The coefficient of J_y gives us the Hall effect, which we need not discuss here. But the coefficient of J_x we should interpret as a resistance, ρ_{xx}. It is the electric field in the *x*-direction, created by unit current flowing in that direction. This is, of course, *transverse* to the magnetic field.

Now for *closed* orbits both σ_{xx} and σ_{yy} behave as A/H^2. The term σ_{xx} in the denominator can be neglected, and

$$\rho_{xx} \sim AR^2. \qquad (10)$$

Thus, the transverse magnetoresistance tends to *saturate* in large fields.

But if there is an *open* orbit along the *x*-direction, σ_{yy} will tend to a constant, B, and both terms in the denominator will decrease as $1/H^2$;

$$\rho_{xx} \sim \frac{1}{A/H^2 + 1/(R^2 H^2 B)} \sim \frac{H^2}{A + 1/R^2 B}. \qquad (11)$$

In this case the magnetoresistance *increases without limit*, proportionately to H^2

The whole effect is dominated by the Hall effect. We do not simply measure the e.m.f., J_x/σ_{xx}, that would normally be needed to make J_x flow. The current J_x gives rise to a Hall e.m.f. RHJ_x in the y-direction. This would cause a current $RHJ_x \sigma_{yy}$ to flow in the y-direction. But the Hall field of *this* current would have magnitude $RH(RHJ_x \sigma_{yy}) = R^2H^2 \sigma_{yy} J_x$ and would appear again as an e.m.f. in the x-direction. This electric field is what we measure and interpret as the e.m.f. associated with the flow of the current J_x through a resistance.

If now we study the transverse magnetoresistance of a single crystal in a very high magnetic field, we may find that there are some directions of H in which $\rho_\perp$ saturates, and others in which it seems to increase indefinitely as H increases. It is clear that the saturation directions must correspond to directions in which all slices of the Fermi surface are closed orbits, whilst the H^2 directions correspond to directions for which some sections contain open orbits. This tells us, at once, that the Fermi surface must be multiply connected. For example, fig. 55 (*a*), the magnetoresistance of Au behaves in a very complicated way

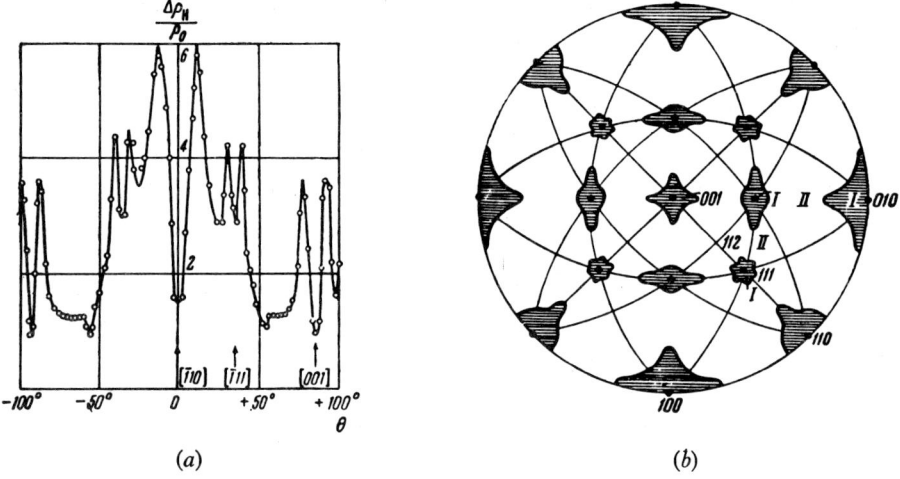

Fig. 55. (*a*) High-field magnetoresistance of Au single crystal (Gaidukov 1959). (*b*) Stereogram of magnetic field directions where transverse magnetoresistance is quadratic in H.

when we vary the direction of the magnetic field. But if we plot on a stereogram all solid angles where the magnetoresistance does not saturate fig. 55 (*b*), we find a relatively simple pattern. Since gold has the same crystal structure as Cu, we look at the possible open and closed orbits on a multiply connected surface like fig. 52 (*c*) as a function of direction, and find that it agrees qualitatively with fig. 55 (*b*). The Fermi surface of Au is thus similar to that of Cu. By studying the shape and size of the non-saturation regions on the stereogram, we can even fix the approximate dimensions of the ' necks '.

High field magnetoresistance is thus a very interesting phenomenon, capable of great use in the detection and analysis of multiply-connected Fermi surfaces, and the establishment of their general topology.

4. Quantization of Orbits: de Haas–van Alphen Effect

A closed orbit has a well-defined cyclotron frequency, ω_H. It is our experience that the energy of any system whose motion is periodic in time can be quantized in units of $\hbar\omega_H$. Suppose that the permitted energy levels must be the cyclotron levels

$$\mathscr{E}_n = (n + \phi)\hbar\omega_H \tag{12}$$

where n is an integer and ϕ is a phase constant—perhaps $\tfrac{1}{2}$. Let us see what this implies.

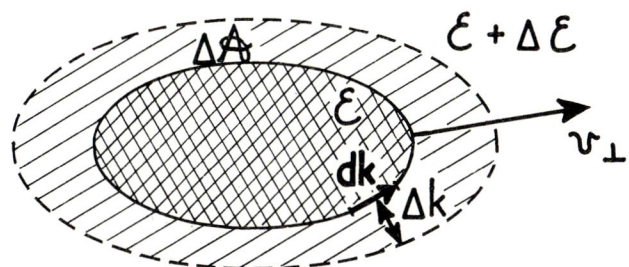

Fig. 56. Geometry of quantization of orbits.

We have already, in eqn. (2), written a formula for the cyclotron frequency:

$$\omega_H = \frac{2\pi eH}{c\hbar} \left[\oint \frac{dk}{v_\perp}\right]^{-1}. \tag{13}$$

And we must remember that v is proportional to the gradient of $\mathscr{E}(\mathbf{k})$ in $\mathbf{k}$-space. Thus, for a small change of energy $\Delta\mathscr{E}$, we move a distance

$$\Delta k_\perp = \Delta\mathscr{E}/\hbar v_\perp \tag{14}$$

outward in $\mathbf{k}$ space, to a new orbit of energy $\mathscr{E} + \Delta\mathscr{E}$ in the plane $k_z = \text{constant}$. The integral in (13) round the orbit is thus proportional to

$$\oint \Delta k_\perp \, dk = \Delta\mathscr{A} \tag{15}$$

—the area in the plane k_z between our two orbits. In fact, from (13), (14) and (15),

$$\Delta\mathscr{A} = \frac{2\pi eH}{c\hbar^2} \cdot \frac{\Delta\mathscr{E}}{\omega_H}. \tag{16}$$

This gives us an elegant formula for the cyclotron frequency:

$$\omega_H = \frac{2\pi eH}{c\hbar^2}\left(\frac{\partial\mathscr{E}}{\partial\mathscr{A}}\right). \tag{17}$$

But it also tells us that if $\mathscr{E}$ is quantized in units of $\hbar\omega_H$, as in (12), then *the area of the orbit is quantized in units of $2\pi eH/c\hbar$*. Thus, for the total area of the orbit of energy $\mathscr{E}_n$

$$\mathscr{A} = (n + \phi)\frac{2\pi eH}{c\hbar}. \tag{18}$$

What we are really saying is that the application of a magnetic field automatically destroys the basic quantization scheme of our electron gas. Except for the component k_z measured along the magnetic field, the components of the momentum $\hbar \mathbf{k}$ are no longer 'good quantum numbers', But the magnetic field introduces new constants of the motion—in particular, the angular momentum about the magnetic field. It is easy to show that this is proportional to the 'area of an orbit in $\mathbf{k}$-space', and is quantized accordingly. Our result (18) is no more than the old Bohr phase-integral formula

$$\oint p\,dq = (n+\phi)\hbar. \tag{19}$$

In the absence of a magnetic field, $\mathscr{E}(\mathbf{k})$ is a simple continuous function of $\mathbf{k}$, with points distributed uniformly in $\mathbf{k}$-space on the grid of 'allowed states'. A magnetic field divides $\mathbf{k}$-space into 'tubes', of constant area of cross-section, corresponding to states of constant magnetic quantum number. Each state changes its energy a little, up or down, so as to condense on the surface of the nearest 'tube'. On the average this process makes little difference to the total energy of the system, since states that have moved up will be nearly compensated by states that have moved down in energy. But consider a sharp Fermi surface,

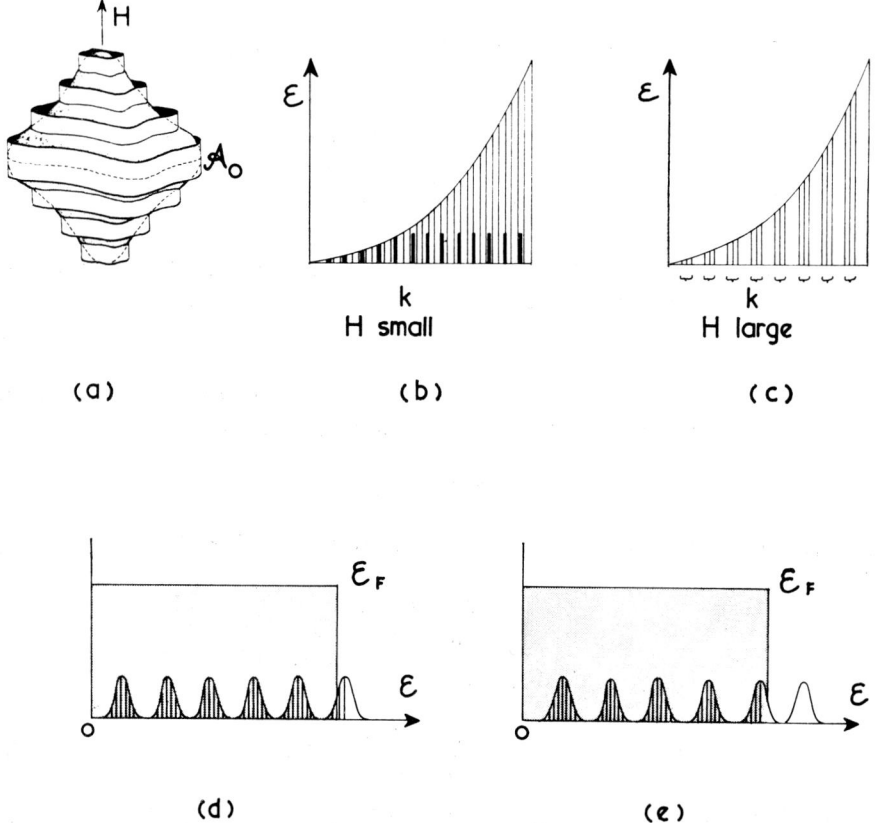

Fig. 57. (a) Cylinders of magnetic levels cutting a Fermi surface. (b) In a small magnetic field, magnetic levels are more closely spaced than the 'allowed states'. (c) In a strong field the states are bunched into magnetic levels. (d) States pulled above Fermi energy as a magnetic level emerges from the Fermi surface. (e) In slightly higher field, a magnetic level can lower the energy of states near the Fermi surface.

as in fig. 57 (d). As we increase the magnetic field, the tubes move outward, because the area of each is proportional to H. Every now and then, a tube will break away from the Fermi surface at a cross-section of maximum area. This is a significant effect; as the tube is drawing away from contact, the energy of the states in this region is being increased, because these levels are still trying to condense on this tube. Then, as the next tube comes towards the surface from below, enough levels become available on it and the energy decreases to below the average value. In other words, there will be an oscillation in the energy of the electron gas, periodic in the quantum number of the tube, of area $\mathscr{A}_0$, touching the maximum cross-section of the Fermi surface. In terms of the magnetic field, the period of the oscillation may be expressed in the form

$$\Delta(1/H) = 2\pi e/\hbar c \mathscr{A}_0. \tag{20}$$

Similar oscillations will also be observed if $\mathscr{A}_0$ is a minimum area of cross-section.

This oscillatory change in energy is detectable as a magnetic susceptibility. It is a contribution to the diamagnetism of the electron gas, since it arises from the 'orbital motion' of the electrons. This is known as the *de Haas-van Alphen* effect. The quantization is also noticeable in other properties of the metal, such as the electrical resistivity (*Shubnikov–de Haas* effect) but the theory is then far more complex and the phenomenon does not seem so easy to detect. The general conditions for the detection of the de Haas-van Alphen effect are, simply, that $\omega_H \tau > 1$, so that collisions should not spoil the definition and quantization of the orbits, and that $\hbar \omega_H > kT$, so that the separation of each tube from the Fermi surface should not be blurred by the spread of thermally excited levels at the top of the Fermi distribution. These conditions can be satisfied fairly easily, especially as the effect can be detected neatly by induction in large, pulsed, magnetic fields, of the order of 10^5 oersted.

As a technique for charting the Fermi surface, the de Haas-van Alphen effect is extremely powerful and useful. The period of oscillation gives directly a simple geometrical parameter of the surface—the maximum or minimum area

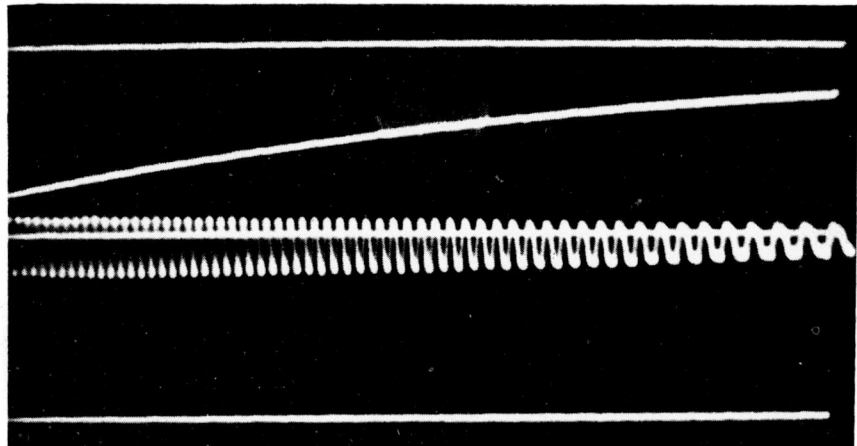

Fig. 58. De Haas-van Alphen oscillations in Cu. The variation of the magnetic field is shown by the smooth curve in the upper part of the photograph. (Shoenberg 1960.)

of cross-section of the Fermi surface normal to the direction of the magnetic field. (It provides us with what are known, I believe, as the 'vital statistics' of the figure.) By making measurements at all orientations of the crystal relative to the magnetic field, one can reconstruct the Fermi surface almost exactly. For example, measurements by Shoenberg on Ag have been fitted to a multiply-connected surface, similar to that of Cu and Au, with a precision of 1 part in 1000 in the radius vector of the surface in any direction. Moreover, the measurement of $\mathscr{A}_0$ is absolute; one can actually check that the Fermi surface contains exactly one electron per atom. This is quite an important point, since it gives an experimental check of the assertion, in III, § 8, that each excitation of the electron gas has the charge, e, of an ordinary isolated electron.

5. Polyvalent metals

In the de Haas-van Alphen effect, the contributions of different parts of the Fermi surface are distinctly visible as different periods of oscillation in $1/H$. This is very important when we come to study polyvalent metals. Not only can we see effects from 'hole' orbits such as the 'dog's bone' in copper, but there may be contributions from several different bands. The procedure for constructing Fermi surfaces in such cases requires some discussion.

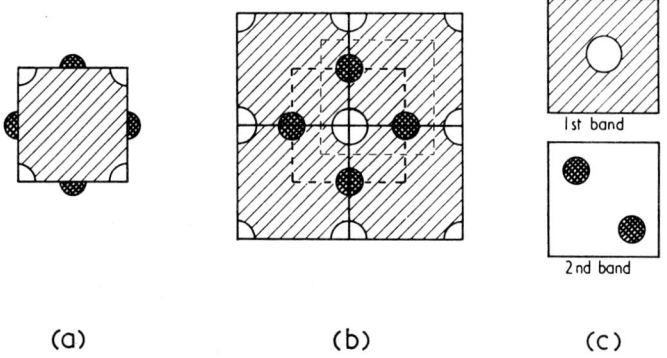

Fig. 59. Schematic Fermi surface for a divalent metal. (a) Extended zone scheme. (b) Repeated zone scheme. (c) Reduced zone scheme, showing 'band of holes' and 'band of electrons'.

Let us, for example consider a hypothetical simple cubic metal with two electrons per atom, with overlaps through the zone faces, fig. 59 (a). According to the rules of zone theory, we are allowed to repeat this figure endlessly. Thus, we get fig. 59 (b). We can here distinguish two 'branches' of the Fermi surface which are not connected by any orbits: in the original zone there are empty pockets which have coalesced to form a closed surface—almost a sphere —of 'holes'. Then there are electrons in the second band that have spilled over the energy gap at the centre of each face. These can be combined into other closed surfaces—three spheres of 'electrons'. It is convenient to say that the Fermi surface consists of a sphere of holes in the first band and three spheres of electrons in the second band. Each of these surfaces contributes separately to the de Haas-van Alphen effect. We often represent the bands separately, as in 59 (c), each inside a single *reduced zone*. It does not matter how we carve this zone out of the repeated zone scheme 59 (b), because the whole figure repeats itself endlessly.

This leads to an amusing geometrical parlour game. Suppose we have a metal such as aluminium, or lead, with 3 or 4 electrons per atom and a complicated polyhedral zone—in these cases the same as copper. We do not know the shape of the Fermi surface. Suppose, though that the electrons were absolutely free. The Fermi surface would be a sphere. Draw this sphere centred on the zone, with the correct volume; it will fall right outside the polyhedron. Thus the first zone will be full.

Now consider the caps of the sphere outside this first zone. These caps will be intersected by zone boundaries—the continuations of the planes forming the faces of the first zone, fig. 60 (a). If the electrons are not precisely free, energy gaps will be introduced over all these planes. Our original sphere will be cut up and we cannot suppose that the magnetic orbits will have the same connectivity as they would for free electrons.

We can easily work out the new topology by repeating the zone. We build up a reciprocal lattice of a number of zones, each with its sphere centred on it, fig. 60 (b). Now look inside any one cell. The spheres centred on other cells will project into this cell—and parts of their surfaces can be pieced together into a connected surface. For example, with three electrons per atom we can construct an object like 60 (c)—a closed surface of 'holes' in the second band.

There still remain some parts of the surface of the original sphere which are not represented in the first two bands. These can be put together into yet another surface inside a zone. We have a multiply-connected surface of electrons, obtained by thickening the edges of the zone polyhedron. In fig. 60 (d) this object is drawn with its zone displaced from the familiar position; it does not matter, as we can draw a cell of the endlessly repeated reciprocal lattice wherever we like in **k**-space. There are also further overlaps right through the corners of the original zone, which can be joined to make small pockets of electrons in the fourth band—rather like the spheres in 59 (c).

For a gas of perfectly free electrons this construction can be made exactly, but has no meaning since there are no energy gaps. In a real metal there are energy gaps, which distort the sphere and separate the bands. But the gaps can only appear on the planes of the zone faces, so that the continuity of the energy function in the zone of each band of fig. 60 will not be changed. The 'limbs' of the 'monster' in fig. 60 (d), for example, might be smoothed out, or might shrink by the withdrawal of electrons into other bands, but they will not be chopped up further. By studying the connectivity of each figure, we may be able to identify the orbits corresponding to the various de Haas-van Alphen periods, or find open orbits to explain the magnetoresistance phenomena. As yet, there has been no analysis of the Fermi surface of a polyvalent metal comparable in certainty and accuracy with the detailed pictures that can be drawn for the noble metals, but this is a programme that is now well under way.

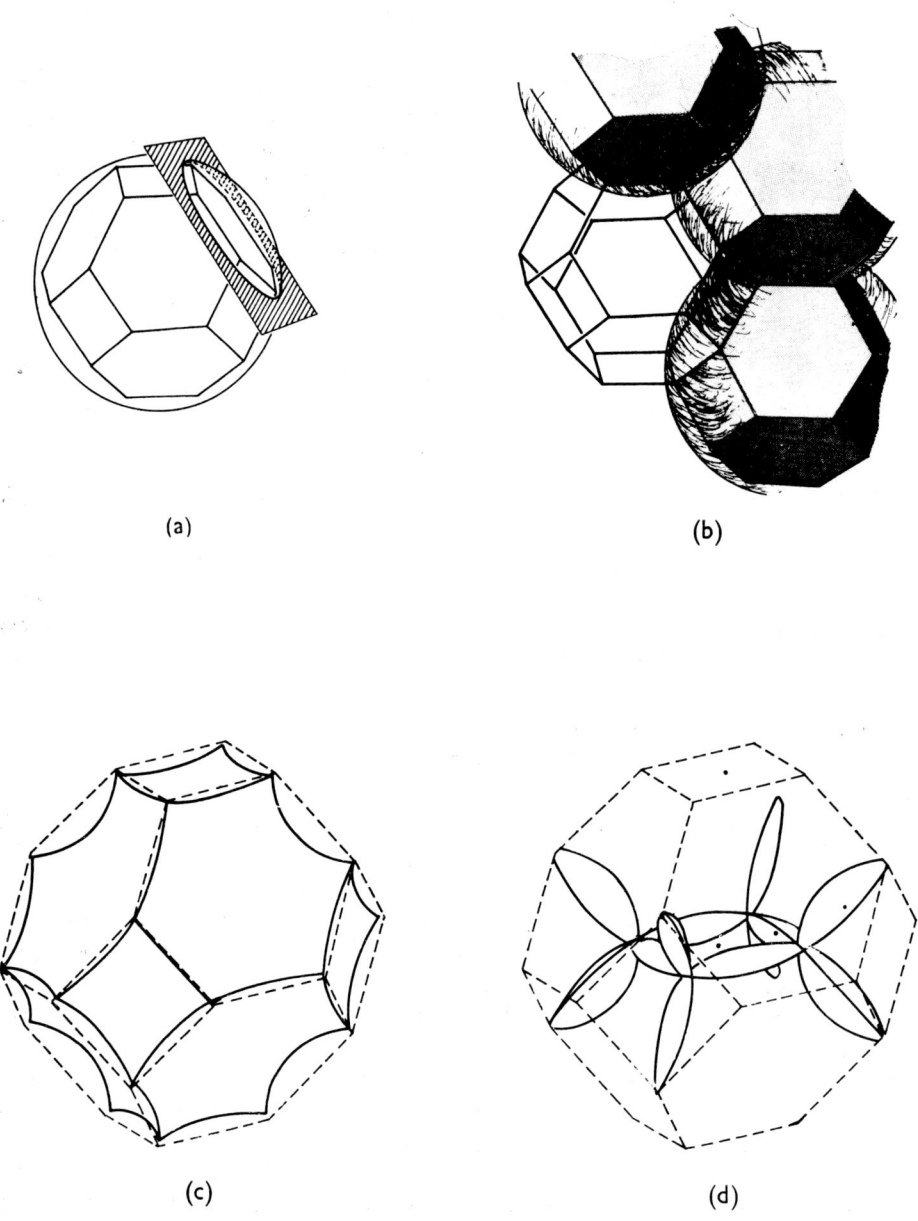

Fig. 60. Reduction of a free-electron sphere to 'zones'. Caps are cut off the sphere, (a) by the boundary planes of the zone which now contains a 'full band of electrons'. To fit the caps together, we think, (b) of enclosing each neighbouring zone of the reciprocal lattice within its sphere: the volume (c) left in the central zone is a 'band of holes'. But there are still some pieces left over, corresponding to regions in (b) where two spheres overlap. These can be fitted together, and moved to the centre of the zone to make (d), a 'monster' of electrons in the third band.

6. Magneto-acoustic effect

We have still not exhausted the list of phenomena which may be exploited in the experimental study of the Fermi surface. By subjecting a specimen of metal to more complex influences we can obtain more complex responses. Suppose, for instance, that we send a high-frequency elastic wave through the crystal. The electrons in the metal tend to scatter ultrasonic waves more or less as if they were 'phonons'. Because of the conditions on energy and crystal momentum (IV, § 5) in this scattering, only certain regions of the Fermi surface contribute to *ultrasonic absorption*. By varying the direction of propagation and polarization of the phonons, one might, in principle, investigate the shape of the Fermi surface.

Unfortunately, this programme is not practical. The theory of the electron-phonon interaction contains unknown quantities, which could vary substantially over the Fermi surface. There is no simple effect that can be interpreted easily in physical terms.

But suppose that we complicate the physical situation still further by applying a magnetic field. A very interesting phenomenon occurs: the coefficient of absorption of the ultrasonic wave oscillates periodically in $1/H$ as we change the magnetic field. The explanation of this effect is as follows:

The magnetic orbit of an electron in **k**-space corresponds to a closed or helical path of the electron in real space. For free electrons the projection of this path on the plane normal to **H** is a circle. In general, we have the equation of motion

$$\dot{\mathbf{k}} = \frac{e}{c\hbar} \mathbf{H} \wedge \dot{\mathbf{r}} \tag{21}$$

for the rate of change of the position vector **r** in real space. It follows at once, by an elementary time integration, that the path in real space, projected on the plane normal to **H**, has the same shape as the 'orbit' in **k**-space, except for multiplication by a factor $(\hbar c/eH)$ and rotation through a right angle about the direction of **H**.

Now pass an ultrasonic wave through the crystal. For example, if **H** is in the z-direction, propagate, in the x-direction, a wave transversely polarized in the y-direction. The lattice will be locally strained, setting up local forces and fields on the electrons. Suppose that at A (fig. 61) this field is parallel

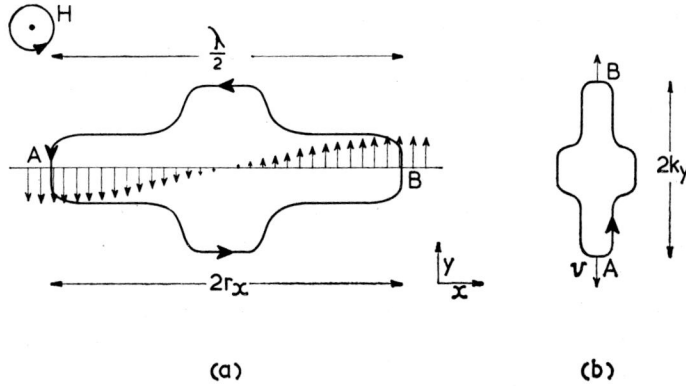

Fig. 61. Magnetic orbits (*a*) in real space; (*b*) on the Fermi surface.

to the velocity of an electron on the Fermi surface. This electron will take up energy from the ultrasonic wave.

If the magnetic field is large enough, the period of a cyclotron circuit can be made much shorter than both the relaxation time for scattering by impurities and the oscillation period of the ultrasonic wave. The electron will make many circuits before it is scattered and in those circuits will sample the electric fields produced by the elastic strain as if they were static. The amount of energy that it picks up will depend on the wavelength of the ultrasonic wave. One can see, from fig. 61, that if there are an odd number of half-waves across the diameter of the path the fields at the extremities will both be accelerating the electron and there will be strong absorption of energy. If there are an even number of half waves, the effects at each end will cancel out. Thus there is a condition

$$2r_x = \frac{\hbar c}{eH} 2 k_y = (n + \tfrac{1}{2}) \lambda \qquad (22)$$

for maximum absorption, with minima at wavelengths in between. Thus, as we vary H, we find oscillations which can be interpreted as a relation between the wavelength, λ, of the ultrasonic wave and the diameter, $2k_y$, of the Fermi surface in the direction normal to H and to the direction of propagation of the wave.

To discuss this effect fully, we need to count contributions from all different positions in space, all sections normal to H, etc. The result is to blunt the oscillations, and shift them a bit. But the attenuation should still be periodic in $(1/H)$ and dependent on an extremal diameter of the Fermi surface in a well-defined direction. These oscillations are, in fact, well marked (fig. 62), both for longitudinally- and transversely- polarized ultrasonic waves and their interpretation, in the case of the noble metals, is consistent with (22). We can, for example, study the dimensions of the 'dog's bone' orbits in Cu, and thus see how much the Fermi surface is pulled out towards the (1,0,0) faces of the zone.

This technique is thus, potentially, very powerful, since it applies a pair of calipers to the Fermi surface and provides the most direct information about its shape. However, one must be a little cautious about this, since the physical situation is much more complicated than in the de Haas-van Alphen effect, and there may be factors (such as the strength of the electron-phonon interaction) which vary so strongly over the Fermi surface that they upset the numerical relations between the 'resonance' field and the extremal cross section.

These techniques have been developed so rapidly in the last five years that one suspects that there may be other methods, even more powerful and convenient, just waiting to be invented, discovered, or applied. One can certainly say that we can now, by deliberate experiment, directly determine the Fermi surface of a metal and thus obtain the most important evidence as to its electronic structure. An international programme (or, if you like, competition) is now in full swing to extend our knowledge from the monovalent metals throughout the periodic table.

Nevertheless, there are still fields in this subject where we are profoundly ignorant. In particular, the electronic structure of alloys is quite uncertain, and the geometrical techniques will not work because the electrons are too strongly scattered. Perhaps, when we fully understand the structure of pure, elementary, single crystals, we may eventually be judged worthy to unravel these more complex secrets of Nature.

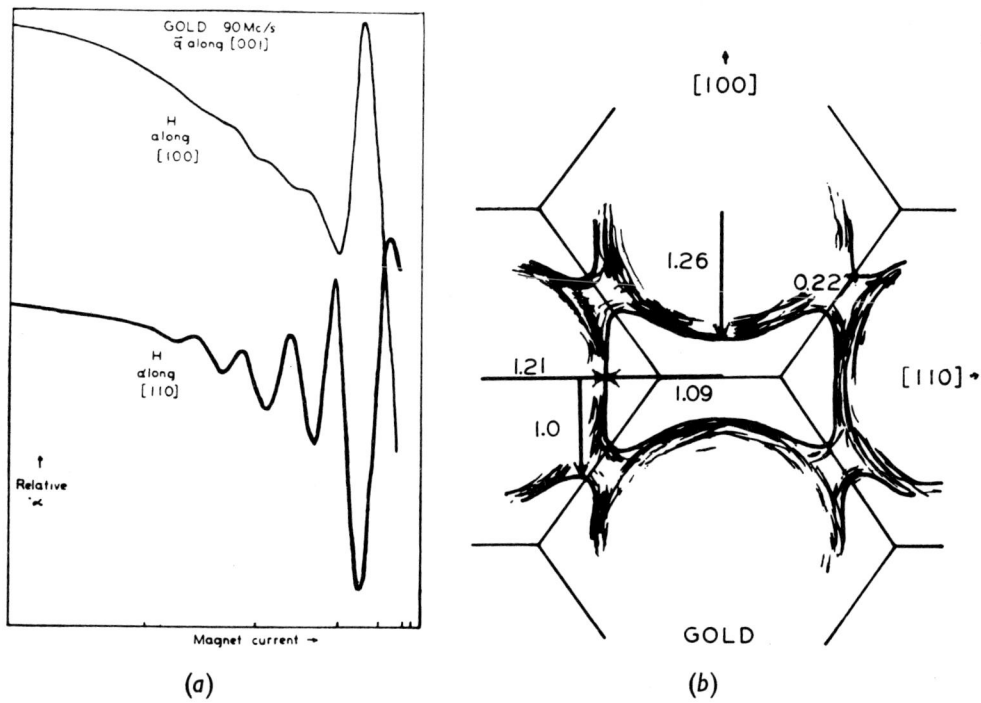

Fig. 62 (a) Magneto-acoustic oscillations in Au (Morse 1960). (b) Dimensions of the Fermi surface from magneto-acoustic data.

Acknowledgements

Figures 47, 53, 55 and 62 are reproduced by permission of the authors and publishers, from *The Fermi Surface*, eds. W. A. Harrison and M. B. Webb (New York: John Wiley and Sons, 1960). Dr. Shoenberg most kindly supplied fig. 58 from his private collection.

References

‡Brillouin, L., 1946, *Wave Propagation in Periodic Structures* (New York: McGraw Hill).
†Harrison, W. A., and Webb, M. B. (eds.), 1960, *The Fermi Surface* (New York: Wiley).
Jones, H., 1960, *Theory of Brillouin Zones* (Amsterdam: North-Holland).
‡Kittel, C., 1956, *An Introduction to Solid State Physics*, 2nd Edition (New York: Wiley).
‡Mott, N. F., and Jones, H., 1936, *Theory of the properties of metals and alloys* (Oxford: Clarendon Press).
Peierls, R., 1955, *Quantum Theory of Solids* (Oxford: Clarendon Press).
†Pippard, A. B., 1960, *Experimental Analysis of the Electronic Structure of Metals—Reports on Progress in Physics XXIII*, p. 176 (London: Physical Society).
Seitz, F., 1940, *Modern Theory of Solids* (New York: McGraw Hill).
‡Slater, J. C., 1951, *Quantum Theory of Matter* (New York: McGraw Hill).

† Basic references for the new techniques for studying the Fermi surface.
‡ Relatively elementary.

SLATER, J. C., 1956, *The Electronic Structure of Solids, Handbuch der Physik XIX* (Berlin: Springer).
WANNIER, G. H., 1959, *Elements of Solid State Theory* (Cambridge: Cambridge University Press).
WILSON, A. H., 1953, *Theory of Metals*, 2nd Edition (Cambridge: Cambridge University Press).
ZIMAN, J. M., 1960, *Electrons and Phonons* (Oxford: Clarendon Press).